Stochastic Communities
A Mathematical Theory of Biodiversity

Stochastic Communities
A Mathematical Theory of Biodiversity

A. K. Dewdney

CRC Press
Taylor & Francis Group
Boca Raton London New York

CRC Press is an imprint of the
Taylor & Francis Group, an **informa** business

CRC Press
Taylor & Francis Group
6000 Broken Sound Parkway NW, Suite 300
Boca Raton, FL 33487-2742

First issued in paperback 2020

© 2017 by Taylor & Francis Group, LLC
CRC Press is an imprint of Taylor & Francis Group, an Informa business

No claim to original U.S. Government works

ISBN 13: 978-0-367-65800-7 (pbk)
ISBN 13: 978-1-138-19702-2 (hbk)

Library of Congress Cataloging-in-Publication Data

Names: Dewdney, A. K., author.
Title: Stochastic communities : a mathematical theory of biodiversity / Alexander K. Dewdney.
Description: Boca Raton : Taylor & Francis, 2017. | Includes bibliographical references and index.
Identifiers: LCCN 2016047383 | ISBN 9781138197022 (hardback : alk. paper)
Subjects: LCSH: Species diversity--Mathematical models | Population biology--Mathematical models. | Stochastic models.
Classification: LCC QH541.15.S64 D49 2017 | DDC 577--dc23
LC record available at https://lccn.loc.gov/2016047383

Visit the Taylor & Francis Web site at
http://www.taylorandfrancis.com

and the CRC Press Web site at
http://www.crcpress.com

Contents

Preface...ix
Acknowledgments..xi
Introduction.. xiii

Chapter 1   The J-Curve and the J Distribution 1

    1.1   The Stochastic Species Hypothesis and the J Distribution 2
    1.2   First Appearance of the J Distribution 4
    1.3   The Community and the Sample.. 6
    1.4   The Need for an Appropriate Theory..................................... 8
    1.5   A Positive Test Using Multisample Data 12
    1.6   Fluctuating Populations.. 14
        1.6.1   The Racetrack Analogy.. 16

Chapter 2   The J Distribution and Its Variations 19

    2.1   Properties of the Probability Density Function..................... 19
        2.1.1   The Capacity Constant 20
        2.1.2   The Mean and Variance of the J Distribution 21
    2.2   The J Distribution: Species and Individuals......................... 22
        2.2.1   The Discrete J Distribution 22
        2.2.2   The Canonical Sequence.. 24
        2.2.3   An Implicit Formula for Rank Abundance 26
        2.2.4   Effect of a Log Transformation on the J Distribution ... 28

Chapter 3   Sampling in Practice and in Theory.................................... 31

    3.1   Randomness and Random Numbers 31
        3.1.1   Generating Random Numbers................................... 32
    3.2   Communities and Samples 34
    3.3   The Sampling Process 34
        3.3.1   The Variety of Sampling Activity 35
        3.3.2   Estimating Sample Intensity 36
        3.3.3   How Samples Have Been Used: Calculating
             Biodiversity 37
    3.4   Computer Simulation of Sampling....................................... 40
        3.4.1   A Sample Simulation Algorithm............................... 41
    3.5   The General Theory of Sampling... 42

Chapter 4   Compiling and Analyzing Field Data 45

   4.1   Histograms and Distributions...45
   4.2   Other Representations: Rank Abundance 47
   4.3   Other Representations: Logarithmic Abundance....................48
   4.4   Estimating Parameters of the J Distribution 49
      4.4.1   The Chi-Square Test...49
      4.4.2   The Kolmogorov-Smirnov Test................................52
   4.5   Application Example: Sample Overlap and Similarity 53

Chapter 5   Predictions from Data 57

   5.1   Predicting Maximum Abundance 57
   5.2   Predicting Species Richness...58
      5.2.1   Counterexample 1...59
      5.2.2   Inadequacies in Current Methods of Estimation 61
         5.2.2.1   The Fisher-Corbet-Williams Method 61
         5.2.2.2   The Goodman Statistic 63
         5.2.2.3   The Jackknife Estimator............................64
         5.2.2.4   The Bootstrap Method 65
      5.2.3   Counterexample 2...65
   5.3   Parametric versus Nonparametric Approaches 66
   5.4   Exact Estimation Methods 67
      5.4.1   The Two-Step Method with an Example....................69
   5.5   Experimental Evaluation of Methods.....................................71
      5.5.1   The Two-Step Method...71
      5.5.2   The One-Step Method...72
   5.6   The Behavior of Error Terms 73
   5.7   Analysis of the Two Sources of Variance...............................74
   5.8   Assessing General Population Declines..................................75

Chapter 6   Extending the Sample..79

   6.1   The Effect of Sampling on Parameters 79
   6.2   Accumulation Curves 81
      6.2.1   Accumulation with Replacement 82
      6.2.2   Accumulation without Replacement 84
         6.2.2.1   Hurlburt's Formula 84
         6.2.2.2   The Hyperbolic Formula 85
   6.3   The Species–Area Relationship...87
      6.3.1   Other Species–Area Laws...89

Chapter 7   Stochastic Systems and the Stochastic Community..........................91

   7.1   The Stochastic Community and Stochastic Systems...............91
      7.1.1   Probability and Time...92
      7.1.2   Generalizations of the Multispecies Logistic System 93
      7.1.3   Stochastic Abundances..95

7.2   Stochastic Communities in Nature 97
        7.2.1   Stochastic Species Hypotheses for Communities 97
        7.2.2   Detecting the J Distribution in Natural Populations.... 99
        7.2.3   The Stochastic Orbit and Its Variations 102
        7.2.4   Long Runs in Stochastic Behavior 104
        7.2.5   Cyclic Changes in Abundance 106

Chapter 8   The Meta-Study: A Review 109

        8.1   Background: The Chi-Square Theorem and Test.................. 109
                8.1.1   An Illustration of the Central Methodology............ 111
        8.2   Applying Chi-Square Theory to Multiple-Source
                Histograms 113
                8.2.1   The Score Conversion Process 114
        8.3   The Meta-Study 116
                8.3.1   The Data Collection Procedure 116
                8.3.2   Testing the Data against Two Distributions 117
                8.3.3   Converting and Compiling the Scores 117
        8.4   Experimental Results of the Meta-Study 118

Chapter 9   Fossil J-Curves 121

        9.1   Background of the Problem.. 121
        9.2   Presence of the J Distribution in Taxonomic Data................ 123
                9.2.1   The Test Method.. 125
                9.2.2   Results of the Study... 126
        9.3   Evolutionary Origin of the J Distribution............................. 128
                9.3.1   Stochastic Genera.. 129
                9.3.2   Episodic Speciation 130
        9.4   Extinction and Speciation in Natural and Artificial
                Communities 132
                9.4.1   Extinction Rates in the MSL System 133
                9.4.2   Speciation in Stochastic Communities.................... 134

Chapter 10   Summary of Theory and Open Problems 137

        10.1   Summary of Research 138
                10.1.1 Examples and Counterexamples 140
        10.2   A Guide to Field Methods and Theory Development 140
        10.3   Open Problems and Prospects.. 141
        10.4   Stochastic Systems as Research Tools................................... 143
                10.4.1 The Stochastic System.. 143
                10.4.2 The Weakly Stochastic System 143
                10.4.3 The Seasonal Stochastic System 144
                10.4.4 The Compartmentalized Trophic System 144
        10.5   Concluding Remarks 144

Contents

Appendix A: Mathematical Notes and Computer Tools 145

Appendix B: Results of the Meta-Study for the J Distribution 165

Appendix C: Results of the Test for the J Distribution in Taxonomic Data... 171

Bibliography 175

Endpaper: Global Map of Biosurvey Sites Used in the Meta-Study 189

Index 191

Preface

In an age when the natural environment is under threat from habitat loss, pollution, and climate change, it is more important than ever that ecologists develop and employ uniform procedures for assessing the condition of all the canaries in the mine, so to speak. Unless results can be compared directly, confusion is likely to result. The lack of a universally accepted (theoretical) species-abundance distribution, along with appropriate methodology for assessment, hampers the field greatly. A widely accepted and well-established distribution would be greatly preferable to the present situation.

The natural focus of the population biology described in the following pages is not the species, but the community of which it forms a part. The behavior of populations may forever defy our efforts to quantify them beyond recognizing their fundamentally stochastic (i.e., random) nature. But a *community* of populations produces a collective behavior, population-wise, that almost always follows the same general pattern, a J-curve. It would seem that unpredictability, far from being a cause for despair, lies at the heart of the exact methods employed here and provides new theoretical horizons for a science that has suffered from a lack of (a) adequate contact with real data and (b) appropriate statistical methods. It would appear that a great deal more data is required for studies of abundance patterns than was formerly thought.

The J distribution described in this monograph appears to capture the ubiquitous shape of the J-curve. Mathematically, it could hardly be simpler, yet it is rather odd statistically, a hyperbola truncated by its axes. This form came as a surprise to me when I found that it was implied mathematically by the central hypothesis of this monograph, the stochastic species hypothesis. Ecologists, field biologists, and other readers of this monograph should have no trouble following the theorems and derivations that appear here, provided they have the minimum expected background of college calculus, probability, and statistics, along with high school algebra. The more involved derivations are left to an appendix, so as to pose minimum interruptions for the reader.

The order in which ideas are presented in this book may call for some patience from the reader. The definitions and applications of the hyperbolic theory are presented in advance of the theoretical arguments and empirical tests that establish the presence of the J distribution in natural communities. For those who wish to reassure themselves by reading ahead, the mathematical derivation of the J distribution from the stochastic species hypothesis will be found in Appendix A, while the empirical meta-study that establishes the presence of the J distribution in field samples of species abundances is the subject of Chapter 8. The stochastic species hypothesis, which leads directly and logically to the (hyperbolic) J distribution, is introduced in Section 1.6 of Chapter 1.

I have been fortunate in my relations with field biologists in assisting with the analysis of many of their datasets, not to mention microbial surveys conducted by me (Dewdney 1996, 2010). My feet, I should hope, are firmly on biological ground.

Crucial to such analyses has been the development of a variety of computer programs, as the reader will discover in the pages to come.

I have also been fortunate to enjoy a profession as mathematician along with some
training in microbiology and an abiding interest in nature in a broad sense. It therefore
seems natural to me to be managing an ATBI (All Taxa Biological Inventory) project
in southern Ontario, even as I work on how populations behave within communities.

Note: In previous publications, I have called the J distribution by another name,
the logistic-J distribution. The latter name has been dropped owing to confusion with
the "logistic distribution" of Balakrishnan (1991), with which it has no particular
relationship. The word "logistic" derived in this case from the limiting effect of finite
biomass on the abundances to be found in any given community of organisms.

Acknowledgments

Special thanks go to those who read an earlier draft manuscript of this book and offered many useful suggestions: Arthur Benke, University of Alabama; Emilio Carral Vilarino, Universidad de Santiago de Compostela; Hsin Chi, National Chung Hsing University, Taiwan; Eungchun Cho, Kentucky State University; Maurice Jansen, Netherlands Ministry of Agriculture; Marcos Mendez Iglesias, Universidad Rey Juan Carlos, Spain; Alfonso Pelli, Universidade Federal do Triangulo Mineiro, Brazil; William D. Taylor, University of Waterloo; and in particular, Dr. Bradford Hawkins at the University of California at Irvine for suggesting use of the Breeding Bird Survey (BBS) database as a more direct test of the stochastic species hypothesis.

I wish to thank the following scientists who made early contributions through discussions and consultations: Robert Bailey, University of Western Ontario; David R. Barton, University of Waterloo; Daniel Botkin, University of California at Santa Barbara; Stanley Caveney, University of Western Ontario; Ryan A. Chisholm, Princeton University; Albert T. Finnamore, Royal Alberta Museum; Kevin Gaston, University of Sheffield; Duncan Golicher, Universidad Autonoma de Mexico; Paul Handford, University of Western Ontario; Rob Hughes, Queen Mary University of London; Tigga Kingston, Boston University; Michael Lynch, University of Waterloo; the late Lynn Margulis, University of Massachusetts; David Schneider, Memorial University of Newfoundland; Gregory Thorn, University of Western Ontario; and Liana Zanette, University of Western Ontario.

I also thank the many sources of support and assistance with (a) the collection of biosurvey literature: (former) graduate students Monica Havelka, Michelle Bowman, Kellie White, Dave Muzia; (b) diagrams and figures: Pablo Jaramillo and Jonathan Dewdney; (c) Adnan Zuberi for equations and formulas; (d) locating literature: Linda Dunn of the University of Western Ontario Libraries; and (e) navigating the ITIS website: David Nicolson of ITIS.

I thank graduate student Gloria Tao, who wrote the software that probed the BBS database and resulted in the change-histogram in Chapter 7. Finally, I am grateful to the University of Western Ontario Biology Department for the Adjunct Appointment and available facilities from 1996 to 2010. During this period, the bulk of the research reported here was completed.

Introduction

...ecologists tend to consume scientific knowledge rather than produce it.

R. H. Peters
A Critique for Ecology, *Cambridge University Press, 1995*

The research described in this book has been aimed at settling the question of how the same general shape of distribution arises when one takes a sample of living communities, whether fish, fowl, or fungi. The shape, variously called the "J-curve" or the "hollow curve," appears in some 15 histograms shown at the end of this introduction, all actual samples drawn *at random* from the literature.

In each histogram, the species are divided into abundance categories. The first or leftmost column counts the number of species having the lowest abundance in the sample. The next column counts the number of species having the next lower abundance in the sample, and so on. Like much biological data, the histograms are neither particularly smooth nor continuous. I have yet to see a field sample where the J-shape does not appear when plotted with appropriate categories. Over the years, some ten formulas have been proposed in the literature as the underlying shape. From the point of view of exact ecology, none of them have been properly tested and none appear to be correct.

Thanks to an accidental discovery made early on, the author now finds himself nearing the end of a 20-year research program. The discovery involved a stochastic simulation of protist predation (Gause 1934) in which the J-curve appeared in the author's computer screen. It led by degrees to a very general scenario called the stochastic species hypothesis wherein the abundances of species in a community fluctuate randomly. The fluctuations could be described as stochastic vibrations based on equal (or nearly equal) birth and death probabilities. A mathematical analysis of the process next revealed a very simple underlying formula called the J distribution. An extensive meta-study involving 125 randomly selected samples from the literature compared the empirical data with the J distribution, revealing a fit (via the chi-square test) that had an overall score that was optimal in the sense that it could hardly be better. The second best proposal had a score that was statistically separable from the optimum, and the remaining proposals were out of the running entirely when it came to closeness of overall fit to the data.

Along the way, a number of useful mathematical theorems have emerged. The most important of these states that the statistical shape of a well-taken sample of a community of organisms must be the same, on average, as the shape of the community itself. In other words, the J-curve that appears in the sample merely reflects a J-curve followed by the community. Given that the sample histogram reflects the J distribution, so must the histogram of the community, albeit with different parameters. This result led directly to a method for estimating with a certain statistical accuracy the species richness of a community based on the richness of the sample. Although several authors have made such attempts, none of the proposed methods carried a reliable error term that could be applied to the estimate.

Other results arose in the course of developing what the author has called "exact methods." Given two field samples from the same general area for example, the underlying species overlap of the samples may be calculated directly with the same degree of statistical precision. In addition, an appropriate species/area formula has been developed, along with other applications.

In 1943, the British ecologist C. B. Williams, working with nocturnal moths flying to a light trap, compiled a great many such histograms and was struck by the fact that, even when different communities of moths were involved, the same pattern inevitably appeared. He called the general pattern the "hollow curve." His opinion about hollow curves lies at the foundation of the research reported here. He thought the hollow curves were essentially hyperbolas. He appears to have been correct.

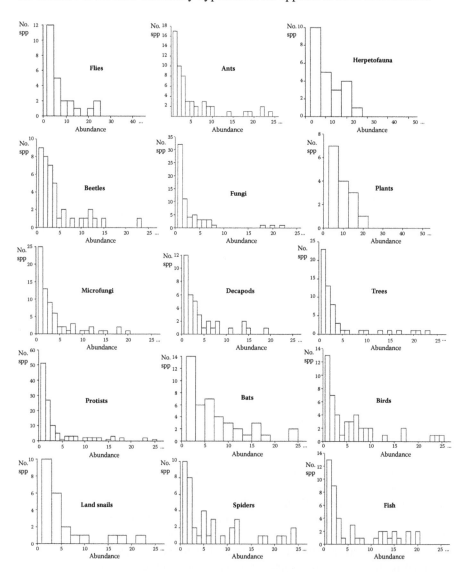

A random sample of random samples hints at the universal presence of the J-curve.

1 The J-Curve and the J Distribution

> Truth proceedeth more readily from error than from confusion.
>
> **Sir Francis Bacon**

Whatever one might mean by the term "biodiversity," the main thrust of all inquiries conducted under that name is the desire to say something meaningful about species of living things in their natural habitat. In particular, the population sizes of various species, whether in relative or absolute terms, become the central focus. The most comprehensive studies address whole communities, namely, all the species within a particular group and living within a specific area, as might be defined by a field biologist. One might wish to know how many species of butterflies there are in a particular forest, for example, along with their relative or absolute abundances. Or one might wish to study a community of aquatic plants in a tropical embayment with the same general aim in mind. Literally thousands of such studies have been conducted for more than 100 years, in almost every clime, habitat, and living kingdom. In all cases—or nearly all cases—the investigator has had recourse to only one portal of inspection, namely, samples collected from or recorded within the community in question.

The main subject of this book is the relationship of the abundances in communities (usually unknown) and the abundances that appear when samples are taken of that community. Short of a Faustian magic mirror that shows us a whole community, along with all its living components, we are stuck with samples. As shown in the illustration in the Introduction, the (so-called) J-curve is ubiquitous in samples of natural communities and there is a clear consensus to that effect, reached in the last decade (McGill et al. 2007). The random sample of a nonnatural community such as the plants in a garden or the animals in a zoo is likely to have a rather different distribution of abundances. The theoretical relationship between the abundances in a community and the corresponding abundances in a sample is described by the Pielou transform (Section 3.5) and is incorporated into a working method for estimating abundances in a community based on one or more samples of it (Section 5.4).

The theory proposed here to account for the ubiquitous J-curve has an associated statistical manifestation previously called the logistic-J distribution (Dewdney 1998a, 2003) but has since been renamed the J distribution, owing to the potential for confusion with the "logistic distribution" (Balakrishnan 1991), with which it has nothing particular in common. Like other proposals to account for this shape, the J distribution is oblivious to the actual species that have a given abundance; only treating the number of species having that abundance as relevant. The distribution is a pure hyperbola,* giving rise to the adjective "hyperbolic," as applied to the theory

---

* The "hyperbolic distribution" (Barndorff-Nielsen 1978) is actually log-hyperbolic.

developed here. The hyperbolic theory has expanded in recent years into a mathematical toolkit for the study of populations in communities that goes well beyond the estimation of abundances.

1.1 THE STOCHASTIC SPECIES HYPOTHESIS AND THE J DISTRIBUTION

The theory described in this monograph begins with the *stochastic species hypothesis* and its logical consequent, the J distribution. In its simplest form, the hypothesis is embodied in a theoretical dynamical system in which a large number of individuals is partitioned into sets called species. Initially, the sets may all have the same size or the size of sets may vary, the final outcome being essentially the same. Individuals are chosen at random from the combined populations and either reproduce or die with equal probability. After each such fundamental operation, the count for the species corresponding to that individual is either incremented or decremented, as appropriate. The hypothesis may be extended in various ways, as in Chapter 7, where it is applied to population changes in communities. In that context, the hypothesis implies a binomial distribution of population changes over time.

As shown in Appendix A.1, a stochastic system produces a specific behavior that we will later claim, on the basis of both theoretical considerations and an extensive study of the sample literature, to manifest as the J distribution of abundances both in communities and in samples of them.

The J distribution is essentially the standard hyperbola, as shown in Figure 1.1, translated in the horizontal direction by the amount ε (epsilon) and in the vertical direction by δ (delta). If we use the variables x' and y' for the primary axis system, the formula for the hyperbola is $y' = \dfrac{1}{x'}$.

The variables x and y for the displaced system obey the following equalities involving parameters ε and δ,

$$x = x' - \varepsilon$$

and

$$y = y' - \delta,$$

and the formula for the probability density function (pdf) becomes

$$f(x) = c\left(\frac{1}{x+\varepsilon} - \delta\right); \quad 0 \le x \le \Delta - \varepsilon$$
$$= 0; \qquad\qquad x \ge \Delta - \varepsilon, \tag{1.1}$$

where $\Delta = 1/\delta$ and c is a normalizing constant (a function of ε and δ) that gives the pdf an area of unity, as required by all probability density functions. This "density" represents the (theoretical) probability of finding a species with an abundance

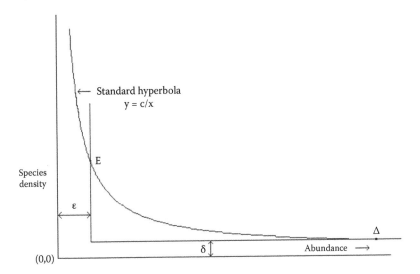

FIGURE 1.1   The hyperbola as the basis of the J distribution.

represented by the area over an infinitesimal interval on the abundance axis, without worrying for the moment about fractional abundances. The constant c has the following formula:

$$c = \frac{1}{\left(\ln\left(\dfrac{\Delta}{\varepsilon}\right) - 1 + \dfrac{\varepsilon}{\Delta} \right)}.$$

In most cases, one may neglect the term ε/Δ, which tends to be rather tiny.

It is interesting that the density function just described is symmetrical under inversion. If one omits the constant c for a moment and writes

$$y = 1/(x+\varepsilon) - \delta,$$

one may express y as a function of x by inverting the function, i.e.,

$$x = 1/(y+\delta) - \varepsilon.$$

The algebraic symmetry merely reflects the geometric symmetry of the hyperbolic function about the diagonal $y' = x'$.

In any case, the J distribution is a two-parameter distribution. The inverses of the constants ε and δ, namely, E and Δ, represent the species peak and the logistic limit, respectively. The first parameter E (when multiplied by Rc, the number of species and the constant c) bounds the number of species in the lowest abundance category, as well as all others. The second parameter Δ bounds the abundance of the largest population. Both "bounds" operate statistically, being simply averages of their respective data. In Appendix A.1, the following elements of the J distribution are placed on a sound mathematical footing as implications of the stochastic species hypothesis:

1. The hyperbolic shape of the distribution is derived in Theorem A.1
2. The parameter δ emerges in Theorem A.2
3. The parameter ε emerges in Section A.1.1

1.2   FIRST APPEARANCE OF THE J DISTRIBUTION

The J distribution first appeared with certainty from a computer simulation written in the mid-1990s to simulate a mutually predatory community of protists. As a long-time sampler of local waters (Dewdney 1996, 2010), I had plenty of microbial data on hand. As a longtime teacher of stochastic simulation and statistics, I had the tools and background necessary to the project.

When it first occurred to me to write such a program, I had already been observing the J-curve in plots of my abundance data, but I had expected the computer simulation to result in something more like a normal curve, with a concentration of species about some mean abundance. Although I was well aware that no microbial community could possibly consist of mutually predatory species of, say, ciliates, it seemed at the time that if any such "community" could produce a unimodal, normal distribution of abundances, it would be this mutually predatory one.

A simulation clock governed the flow of events in the system; at each tick of the clock, two individual "organisms" would be selected from the total community population. Without having to represent the organisms at all, the simulation would simply ensure that the species of the first individual had its population incremented by 1, the species-token moving, in consequence, one unit to the right along the abundance axis. Meanwhile, the species of the second individual would have its population decremented, with its species shifting one position to the left. In such a system, the total number of individuals would be preserved by the "predation" operation.

The original purpose of the simulation was to demonstrate that such a community would achieve a balance of populations, with all species fluctuating about the same mean population size. This did not happen. Indeed, a J-shaped curve stubbornly emerged every time the simulation was run! Each run would begin with a single spike, with all populations occupying the same abundance category. Over time, the spike would flatten into a skewed, bell-shaped curve, where I had expected the process essentially to stabilize. Instead, the process continued as the curve continued to flatten out, even as some species piled up at the low end of the abundance scale while others fled randomly to the high end. In most of these mini-experiments, I kept the average abundance low in order to encourage a more definite shape. Accordingly, I kept the extirpation switch in the "off" position in order to allow the shape to build. (Descriptions of this and other systems used in this research are found in Appendix A.3.)

During the equilibrium phase of the simulation, histograms with a low mean abundance bore a close similarity to the field samples that I had already begun to collect from biosurveys of other kinds of communities. Indeed, they were (visually at least) indistinguishable from them. Yet only a small fraction of the samples I had been examining involved predators of any kind. It took a surprising amount of time (in retrospect) to realize that the simulation was not about predation, with one organism dying while another reproduced in consequence. It was about the birth/death processes itself, in particular the equality of probabilities of births versus deaths. After all, there was no actual predation in the dynamical system. Whenever one species moved to the right, some other species would move to the left. It was a clear case of a specific model turning out to be rather general, after all.

An initial result that appeared to confirm the presence of a hyperbolic function relied on the analysis of a slightly more general system in which a random individual was selected from the set of all individuals and either duplicated or deleted, both events taking place with equal probability (see Appendix A.1 for the Equilibrium Theorem). After a year of studying the more general system, I obtained a more exact result: the equilibrium solution of the mathematical system was indeed a hyperbola, but translated downward by a small amount that I decided to call "delta." The formula was

$$k\left(\frac{1}{x} - \delta\right),$$

where k is a constant (Dewdney 1998a). New simulation programs were written to explore the idea of equiprobable birth/death processes under varying conditions on the balance of probabilities governing them. The probability of birth, for example, could exceed that of death for a time, but it would eventually move in the other direction—all at random. These more general versions of the dynamical system, referred to generically as multispecies logistical systems, produced the same J-shaped curves as the earlier system did. Indeed, they were indistinguishable from the histograms of field surveys that I had already been collecting. One could hardly be blamed for suspecting that the J-curve was not only a universal phenomenon, but the reflection of a simple equilibrium process.

In the new simulations, each individual "organism" had an approximately equal probability of dying or reproducing at each moment. Populations fluctuated randomly, but ultimately, a few became larger, while many became smaller. The result, as in the case of the predation system, was distinctly counterintuitive and illustrates the dangers of armchair theorizing. At equilibrium, which never failed to develop, the histogram would be frozen and compared with a hyperbola (visually). The fit was often rather good, taking into account the usual statistical fluctuations. Since the net effect of the births and deaths in this program was to preserve the total "biomass" of the system, the logistic limit could be applied and the same formula derived from the predation system applied here, as well. At this point, a general mechanism suggested itself. Called the stochastic species hypothesis, it postulated equal probabilities of birth and death over a fixed, relatively short period of time, with corresponding changes in population thus expected. The hypothesis is described and explored in some detail in Section 7.1.2.

Following an intensive literature review of other species-abundance models, I was surprised to find that none arose from what might be called systems considerations. But what about the data itself? Did the J distribution actually show up in field data? A massive study was launched. With the aid of graduate students in our Department of Biology, at the University of Western Ontario, I began a random collection of biosurvey papers, ensuring that all the major classes of biota were covered as they went. Each of the resulting histograms was compared with a version of the J distribution that shared the same mean and height of initial peak; given values for these quantities, the values of the parameters ε and Δ are uniquely determined. (See the Transfer Equations in Appendix A.2.8.) The resulting chi-square scores were normalized to 10 degrees of freedom to make cross-comparisons possible. (See the scores histogram in Figure 1.3.) The average score that emerged from the study was 10.43, very close to the optimum score of 10.0 and well separated from the average score that emerged from parallel tests of the distribution that most closely resembled the hyperbolic curve; the log-series distribution (see Section 1.4) scored over 13. It was unnecessary to test other distributions, since they resembled the J distribution even less and were sure to yield even higher scores. Only something with a very similar shape could hope to score as well (Dewdney 2003). In short, there is no "room" for another distribution in view of the optimality of the scores. The results of this meta-study amounted to what can only be called "strong support" of the stochastic species hypothesis and, therefore, of the J distribution that appears to emerge from it as a universal natural phenomenon.

1.3   THE COMMUNITY AND THE SAMPLE

Figure 1.2 illustrates the relationship between a natural community and its sample. Each little square represents a species, its position on the horizontal axis representing its abundance. It will be a general feature of the histograms shown in this book that the numerical label associated with each abundance appears at the right-hand side of the corresponding interval on the abundance axis. This is a technical device that harmonizes discrete and continuous versions of the J distribution. Some theoretical abundances, as well as some empirical ones, are fractional, with values such as 3.2, 3.7, and 3.9, for example, all gathered into the category 4.0. With some exceptions the kth category would embrace all abundances lying in the half-open interval, $(k - 1, k]$.

It will be a common experience for readers of this book to encounter irregular histograms like the ones in Figure 1.2 where, due to horizontal space limitations, not all species are shown. In cases where it matters, the missing abundances are always given. In cases where it does not, the reader may imagine them as being present, nevertheless. Indeed, the J distribution indicates just how far out on the abundance axis such species might be expected to occur.

The histograms shown in Figure 1.2 are imaginary but reflect what one might call realistic variability. Those with experience of actual sample histograms may well accept the sample data as typical, but who has seen the histogram of an entire community? The community histogram is simply an indication of what a community might look like, given the nature of the sampling process. The horrifying

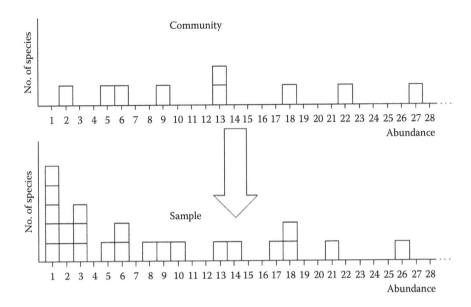

FIGURE 1.2 The relationship between abundances in a community and those of a sample.

sight presented in the sample histogram of so many species seemingly about to be extirpated vanishes at the sight of the community histogram, with such a peak being absent or weakly developed. In any case, from the point of view of the present theory, it would be misleading to describe the peaks as "anomalous" (Coddington et al. 2009). As I hope to demonstrate in this monograph, they are entirely natural. It must be remembered that, in general, the abundance of a species in a sample is usually much lower than its abundance in the community from which the sample was taken. For example, a species of abundance 1 in the sample histogram might well have abundance 27 in the source community. The sampling process has "mapped" the higher community abundance into one of relatively low abundance.

Although the focus of this work is ultimately the community under investigation, we have only samples to work with. As a result, most of the histograms presented in this book belong to either hypothetical or actual samples. As shown in the illustration in the Introduction, they are generally characterized by a high initial peak that is followed by a rapid descent in bar height, whether smooth or ragged. The descent becomes more gradual with ever higher abundances, there being ever fewer species at these abundances. The theoretical J-curves that represent these histograms have the same general shape, although they have a smoother appearance.

The following sections in this introductory chapter include a brief review of other proposals for abundance distributions, along with an explanation of goodness-of-fit tests and their use in both negative and positive mode. An introduction to the J distribution beginning with the probability density function (pdf), an explanation of the two associated parameters, and a formula for the mean μ will appear in the next chapter.

1.4   THE NEED FOR AN APPROPRIATE THEORY

It is an unusual development in any science that a plethora of formulas should result from the attempt to capture a natural phenomenon in mathematical terms. In the normal course of research any formula found wanting would be discarded and a new formula developed. Since the early 1940s, about a dozen formulas for species abundances (in samples) have been proposed, with none emerging as a winner, so to speak. Here are the principal ones discussed by Magurran (1988). She also mentions another three proposals, for a total of seven.

1. The geometric series (May 1975)
2. The log-series distribution (Fisher, Corbet, and Williams 1943)
3. The lognormal distribution (Preston 1948) (see Sections 3.6 and 3.6.1)
4. The broken stick model (MacArthur 1957)

Among the other proposals discussed by Magurran, special mention should be made of the dynamical model of Hughes (1986), the only proposal prior to the 1990s to be tested against an appropriately large number of samples from the field. Unfortunately, Hughes was never able to find a closed formula for the J-shaped curves that his data produced.

It is difficult to characterize succinctly, without committing the sin of oversimplification, how ecologists interested in theory have worked in the past. However, it is fair to say that the various proposals just indicated have been compared with various sets of field data over the years with the aim of discovering which proposals fit the field data best. This seems like a perfectly logical procedure, but it contains a very dangerous pitfall in the form of an unstated (and apparently unrealized) assumption. From the point of view of statistical theory, the outcome of this method is (when one thinks about it) predictable, with each distribution fitting best with certain datasets and not fitting others as well. For example, four of the proposals listed previously are described by Magurran, in her research summary, as follows. "Based on the rank-abundance representation of field data" (Section 4.2 of this book), "the geometric series best fitted" (apparently by visual inspection) "plant data from a sub-alpine forest, the log-series and the lognormal best fitted a histogram of plant data from a deciduous forest, and the broken stick model best fit a survey of nesting birds in a deciduous forest."

The methodology of comparing a theoretical proposal with a handful of field histograms has persisted into the new millennium. For example, Hubbell (2001) continues this methodology and, in a sense, brings it to a climax, drawing support for various manifestations of his zero-sum multinomial distribution from seven studies of various biotic groups, from trees to birds and, not surprisingly, finds a good (visual) fit within a milieu that contains previously proposed distributions implicitly. Connolly et al. (2005) compared four different distributions with survey data involving corals and fish in Pacific Ocean reefs, finding that some distributions fit the data from some reefs better than others do, where other distributions did better.

As more and more datasets are compared with the limited number of distribution proposals under active consideration, it is not surprising that each camp builds up followers, so to speak. A summary of the situation provided by Deserud and Engen (2000) points out that "A large number of data sets from ecological communities correspond well to the model in which the abundances are lognormally distributed…" later pointing out in the same breath that "lognormal models may often provide a rather bad fit to observed data."

As if it were a tacit admission that something is wrong, there has been a long tradition in theoretical ecology of attempting to unify somehow the various distribution proposals into one. It is not difficult to find mathematical generalizations of the proposals mentioned earlier. But was it the unsatisfactory results of such unification projects that have spawned so many attempts? Beginning with Sugihara (1980) and culminating with Deserud and Engen (2000), then Hubbell (2001), the unification process has served to keep alive the idea that each proposal has its niche, so to speak. More recently, a group of 18 authors (McGill et al. 2007) have proposed "moving beyond single prediction theories to integration within an ecological framework." The article demonstrates clearly a continuing confusion about the roles of species abundance distributions in an ecological context. It is unclear how to "move beyond" the present situation until it is corrected.

The underlying methodological weakness has resulted in no little soul-searching, as when a year later, the 14-author paper by Doak et al. (2008) declares that, in view of science being full of surprises, "…it is not so surprising, so to speak, that we frequently face outcomes of experiments and observations that leave us scratching our heads, wondering how we could have been so wrong in our expectations." In the context of the methodological problems just described, it is not surprising at all.

Two main assumptions appear to support the traditional approach to the assessment of distribution proposals. It is not clear from my reading of the literature that either assumption has ever been explicitly stated in some journal article, but most would agree that they would be necessary assumptions for the method just described to work:

a. The shape of the sample histogram, however its abundances are plotted, reflects in a specific way the shape of the community.
b. The shape of a community histogram is stable in the sense that two samples that are well separated in time or space will reflect that community.

The first assumption is an article of faith with field biologists. If the samples they so painstakingly gathered in the wild turned out to have no such relation with the community, there could only be despair. In fact, the distribution underlying the sample "reflects" the distribution that underlies the community in a statistical manner. The underlying distributions are the same, on average, but for a change in parameter values (Dewdney 1998b). Thus, whatever one may mean by the (deliberately vague) word "reflect," the first assumption is completely solid.

A reasonable abstract description of any "community" can be framed in the context of the upper histogram in Figure 1.2. As I have pointed out, the histogram of a community would have a rather strung-out appearance, in comparison with the histogram of its sample. Over time, the various populations that make up the community will all fluctuate, some increasing, some decreasing, with occasional changes in direction, seemingly at random, among all of the populations. In terms of the histogram itself, one may visualize these fluctuations as a sort of vibration when the film is speeded up, so to speak. Species at the high abundance end vibrate at a higher frequency because their populations are much larger, with births and deaths occurring at a higher rate over time. At the low abundance end, in corresponding fashion, the species move in more sedate fashion, slowing rather quickly as they approach extirpation.

The reasons for the apparent randomness in the motion of these species are typically myriad. Suffice it to say that the overall shape of the histogram, however one described it, would also change accordingly. All proposals for the underlying theoretical distribution share what might be called the J-attribute (albeit in a mild form): an initial peak is followed by a more or less smooth decline. All samples from the field also share the J-attribute, albeit more strongly peaked and with a steeper initial decline. Within such a descriptive framework, one may find room for any of the extant theoretical proposals. Samples of the putative community might well, over time, exhibit the full range of these shapes. Would they change enough to defy a description as "geometric," "log-series," "log-normal," or "broken stick"? The answer is simple: They would change their overall shape enough, over time, to defy any proposal, including the centerpiece of this book, the J distribution.

I cannot account for the reasoning behind Assumption b, but it may have something to do with the notion that a large sample size guarantees statistical significance. In other words, the large size of the field samples that have been used to promote one distribution over another were themselves thought to guarantee that the overall results were somehow determinative. Assumption b amounts to an error in reasoning that I shall call the *error of misplaced generality*.

According to the first assumption, as a community distribution changes its shape over time, so would the samples taken of it. Studies that would attach significance to such changes (Magurran 2007) fall into essentially the same trap. Changes in the shape of a community distribution will happen willy-nilly, and the new variations of the J-curve, as revealed (or not) by the sample distribution, have little actual significance. The hyperbolic theory proposed in this monograph tells us that living communities in general have no particular shape beyond displaying the J-attribute. This is true of virtually all of the biosurveys used in the meta-study that is described in Chapter 8. Yet, their net effect, a special kind of average, points to the J distribution rather convincingly.

The sub-alpine community of forest plants that supported the geometric distribution today may well betray it utterly 50 years later, favoring the log-series distribution instead, perhaps. With this view of natural communities in mind, conclusions about proposed distributions arrived through misplaced generality are best regarded as coincidences. Made 30 years earlier on corresponding data, the order of the four distributions might well be scrambled. In view of this unfortunate methodology, one could substitute any four distributions one liked—made up for the occasion—then rewind the historical tape and watch very much the same papers appear, favoring one distribution or another or perhaps attempting to unify them! To put the same point more precisely, it is certainly possible to submit all 125 datasets in the meta-study reported in Chapter 8 to a simple test. It would be a bit labor-intensive, but one could compare all 125 datasets with each of the theoretical distributions that have appeared in the literature to date. It is an elementary observation that every proposal will have some datasets fit that distribution better than any other. Some will have more than others, of course, and judging from the results of the meta-study, the J distribution would have the most.

In the remainder of this chapter and in the chapters to come, I will argue that there can be only one universal statistical descriptor of abundances in natural communities.

As for other problems with the methodology just described, one might also point out that goodness-of-fit tests that are standard in other fields were rarely employed in the assessment of these shapes, with only two such tests having shown up in a context of several hundred articles. Tests such as the chi-square or Kolmogorov-Smirnov take two histograms and compare them, producing a numerical measure of the degree of the similarity. In a field that yearns for quantitative precision, why would anyone neglect a perfectly good numerical measure of similarity between theoretical and empirical histograms? The fact that such measures of similarity have not been used may explain why there is so little awareness among theorists of the enormous variation to be expected from a single underlying distribution. The chi-square curve (see Figure 1.3) has a long tail to tell, so to speak, of the many field histograms that will fit rather poorly without failing to arise from the distribution under test.

As if to make up for the lack of quantitative measure, Akaike information criterion (AIC) methods have come into increasing use in the new millennium. The criterion is embodied in the formula

$$2k - 2\log(L),$$

where k is the number of parameters in the model being compared to a sample and L is the maximum likelihood estimator (MLE) for the variable being measured (Akaike 1974, Burnham and Anderson 2002; see also Schwartz 2011 in web references). Typically, the criterion is used with several datasets at a time, the one having the lowest AIC score being judged closest to the model in question; although easy to apply, the criterion works in a purely comparative milieu, telling the researcher only which of several sets of data is closest to the model. In the case of a distribution in standard form (species per abundance), one would use the MLE derived from the theoretical distribution formula for each abundance category taken separately and then take a weighted average of numbers so derived.

The AIC measure, far from being an improvement on the general disuse of similarity measures as described earlier, actually makes the situation worse because it does nothing to dispel the illusion generated by the error of misplaced generality.

To someone who understands the error, the following little story is not entirely unfair: a child was found one day with three buckets into which he was busily sorting pennies. The buckets were labeled "unlucky," "normal," and "lucky." He would take a penny from a pile beside him and flip it 10 times, counting the number of heads, then placing the penny in the "lucky" bucket if it came up heads more than seven times. If it came up heads fewer than four times, he would deposit it in the "unlucky" bucket. The remainder ended up in the bucket labeled "normal." The specific under-lying source of a sample no more inheres in a single field histogram than the element of luck inheres in a penny. In this little tale, the role of the J distribution is played by a single number, the overall probability *p* of 0.5.

Given the difficulties just described, it may be suspected that the field of theo-retical ecology is in the grip of an unacknowledged crisis. The failure to appreci-ate the statistical behavior of abundances in communities has become a nursery of untested proposals. The crisis is only made worse by the adoption by some ecologists of a social constructivist ethic, as enunciated by Hilborn and Mangel (1997) in *The Ecological Detective*. In this context, "...there is no correct model." Ironically, the statement was largely true up to the time of that book's publication. The notion that all "models" are more or less acceptable is nevertheless unhelpful. At this point, the quote from Sir Francis Bacon at the head of this chapter comes into play: If, in a given field of inquiry, it is not possible to be in error, then it is not possible for that field to be a science. After all, that is exactly what the quote means.

1.5   A POSITIVE TEST USING MULTISAMPLE DATA

As normally applied, goodness-of-fit tests are used to reject hypotheses. As such, the rejection, when supported by the test, applies to each sample tested. Nonrejection of a particular fit does not amount to "acceptance," as such, since the tests are not sym-metrical in this respect. Nonrejection, when based on a single sample or just a few, may be taken as evidence in favor of the null hypothesis, but as "evidence," it is weak and cannot be used, by itself, to establish anything. Since such tests are used here in affirmative rather than rejective mode, a great many samples, rather than just a few, are needed to establish the presence of an underlying distribution.

This last point is important enough to expand upon so that the reasons for the multitest requirement are made plain. As normally used in curve-fitting applications, a test such as the chi-square compiles the chi-square statistic or "score" as follows:

$$\chi^2 \text{ score} = \sum \frac{(t_i - e_i)^2}{t_i}.$$

The index i ranges through abundance classes 1, 2, 3, etc., and the variables t_i and e_i represent the number of species having abundance i in the theoretical and empirical distributions, respectively. The greater the difference between the theoreti-cal prediction (t_i) and the empirical datum (e_i) is, the greater is the contribution to the previous statistic. Thus, higher chi-square scores reflect a rather poor fit to the data, while lower scores represent better fits. The same thing is true of the Kolmogorov-Smirnov test.

Used in the normal way (Hays and Winkler 1971), a goodness-of-fit test will determine the likelihood that a given set of empirical data *does not have* a particu-lar theoretical distribution. In such a case, a chi-square score, for example, can be used to reject the distribution if it exceeds a threshold or critical value, as given in a standard chi-square table (see Chapter 4). In such a case, the null hypothesis (that the empirical data follow a particular theoretical distribution) is rejected. In normal parlance, the hypothesis is said to be "accepted" if the score is at or below the appropriate critical value. As already pointed out, such terminology does not actually mean that the data in question have the distribution under test. After all, there are infinitely many curves, each with a different mathematical formula, that would fit the data as well or better than the distribution under test. How could a particular form be confirmed as the underlying distribution?

To use a goodness-of-fit test in confirmatory mode, one must have a great many datasets (say 100) and one must perform a goodness-of-fit test on all of them, compiling the scores themselves into a histogram that may be compared directly with the chi-square distribution itself, as in Figure 1.3. The figure shows the chi-square distribution (smooth curve) with the histogram of 125 test scores superimposed upon it. The scores are the result of the meta-study reported in Chapter 7 of this monograph. The height of bar above a given numerical category represents the number of chi-square scores that fell within that category. Thus, four scores with values in the range (4.00, 5.00] happened to fall in category 5.

The theoretical chi-square curve clearly reflects the expectation that some samples will have very poor (i.e., large) scores, while others have very good ones. Indeed, the actual scores thus achieved largely fulfill this prediction. It is a key feature of chi-square theory that if a great many such scores emerge from a single underlying distribution, their average value will equal the number of degrees of freedom of the tests themselves. If all the tests were conducted at 10 degrees of freedom, for example, the average expected score would be 10.0 or very close to it. The chi-square envelope in the figure is based on 10 degrees of freedom, while scores emerging from the meta-study were normalized to 10 degrees of freedom. The resulting average score of 10.43 is somewhat greater than 10, the score bars being collectively shifted slightly to the right. In any event, it is not possible for results to be shifted to the left by any significant amount, since this

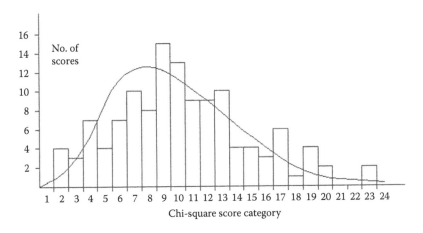

FIGURE 1.3 Distribution of chi-square scores: bar height = number of scores.

would violate Pearson's theorem. The overall fits were optimal in this sense. If other proposed distributions have test scores that are far enough away from the optimal mean to be statistically separable from it, the J distribution is clearly preferable (Dewdney 2000).

A comparison of Figure 1.3 with Figure 8.3 in Section 8.1.1 demonstrates clearly the nature of such fits. The histogram in Figure 8.3 was derived from an exact experiment in which *all* the simulated samples came from exactly the same (uniform) distribution. Yet, when it comes to a visual comparison of histogram and the accompanying theoretical curve, particularly the ragged appearance of both, there is little to choose between the two figures.

Among all other proposals for a theoretical abundance distribution, the distribution closest to the J distribution in overall shape is the log-series distribution. When put through the same series of tests against empirical data, the log-series had a significantly higher score of 13.56, overall. Its chi-square "envelope" would overlap the one shown in Figure 1.2, but it would be displaced a few categories to the right. Other distributions proposed for the role of species/abundance descriptors, such as the lognormal distribution, resemble the J distribution far less, and one does not need to subject them to the same test since their scores would be far too high to be in the running. To do as well as (or better than) the J distribution, a proposed alternative must resemble it very closely. I will return to the meta-study in Chapter 8, providing all the detail necessary for its evaluation as a research tool.

It is an interesting historical fact that the British Biologist C. B. Williams thought that the "hollow curves" he was seeing in lepidopteran light-trap data collectively resembled hyperbolas (Williams 1964). He appears to have been right. His statistical colleague, R. A. Fisher, talked Williams out of the hyperbola, declaring that it was unsuitable for use as a statistical distribution, since it had a nonfinite area under it. Fisher was also right, but the idea of a translated hyperbola truncated by its axes obviously never occurred to him. Given that the subtleties of randomly fluctuating populations would not become apparent until computers were widely available, Fisher can hardly be blamed for this (see next section). In any event, it was not Fisher who developed goodness-of-fit tests, but his archrival (statistically speaking), Karl Pearson. Fisher undoubtedly knew about goodness-of-fit tests but seems to have avoided them for some reason. Ironically, in spite of Willams's insight, it may have been here that the pattern of inappropriate testing in theoretical ecology first developed.

The foregoing provides some background to the development of the J distribution that may be useful in clarifying its role in species-abundance studies. In the next chapter, I will provide more of the mathematical details connected with the J distribution, and in subsequent chapters, I will show how the distribution may be used in the field to make reliable statistical inferences about communities from their samples.

1.6 FLUCTUATING POPULATIONS

The stochastic species hypothesis (see Section 7.1.1 for a full description) might seem, at first sight, to justify a great throwing up of the hands: if it is all random, what is the point of trying to tease out ecological mechanisms to account for population changes? The answer is simply that the stochastic species hypothesis provides a framework of understanding within which such studies may be pursued. Populations fluctuate for a huge variety of reasons, each with its own mechanism. The net effect of these mechanisms is "effectively random" (cf., Chapter 4) even though one mechanism may dominate during one period of time, another later.

Randomness in natural populations has long been suspected but has faced an uphill struggle in the academic forum against deterministic ideas, which can be classified into a continuum. The first view is that all species maintain more or less fixed populations (Marsh 1865), a common nineteenth century understanding of nature. Amazingly, this view persisted to some degree throughout most of the twentieth century. For example, in a book on population genetics, Wallace (1981) declares more than 100 years later, "The second thing we can say about populations is that, despite temporal fluctuations, in the long run they tend to remain constant in size."

This "common understanding" was replaced, at least for some species of animals, by the notion of predator-prey cycles, as predicted by the Lotke-Volterra equations (Leigh 1968). Backed by famous datasets such as the Hudson's Bay trapping data (Odum 1953), the theory gained wide acceptance and is still believed by many ecologists today, in spite of experimental evidence that such cycles do not always develop, e.g., the Gause experiments (Gause 1934), as described in Botkin (1990). However, a more modern view, that of "density regulated" populations, held that populations did, indeed, fluctuate randomly, but only within limits imposed by the density (relative abundance) of a species (Smith 1935). Despite the extensive literature on density regulation that has developed since Smith's time, this concept of population behavior is not universally accepted. Thanks in part to the popularity of chaotic population dynamics (May 1974) and in part to the failure of the population-regulation school to come to firm conclusions (Cappuccino and Price 1995), there would appear to be a growing suspicion that fluctuations in natural populations are indeed random, the same view pursued by Hubbell (2001). An interesting example of accommodation between the two views addresses the concept of "density-vague" behavior, in which populations are bounded away only from extreme density and extirpation (Strong 1986). Not surprisingly, the literature is sprinkled with negative results, as in the comparison of a great many population datasets with random walks that yielded very little difference (Den Boer 1991).

About the problems of coming to grips with population fluctuations, Botkin (1990) writes:

> At the heart of the issues are ideas of stability, constancy, and balance, ideas intimately entwined in theories about nature. Perhaps one reason that the deficiencies of the theories were not examined or tested adequately by observation in the field—out in nature—was that ecologists were typically uncomfortable with theory and theoreticians. Doing science and creating theory were commonly distinguished as separate activities. Although theory was typically considered not to be necessary or important to the practicing ecologist,... theory played a dominant role in shaping the very character of inquiry and conclusions about populations and ecosystems (i.e., about nature). As Kenneth Watt wrote in 1962, ecologists had tended to believe that their science had lacked theory, while in fact it had 'too much' theory—in the sense that the theory

had been utilized and was influential even though it was not carefully connected to observations. 'Field ecologists,' those making measurements and observations in the forest and field, generally did not understand the mathematics of the logistic and of the Lotka-Volterra equations. But since physicists and mathematicians had the highest status among scientists, and since what physicists and mathematicians generally said was generally right, field ecologists tended to regard the logistic and the Lotka-Volterra equations as true. Lacking the understanding to analyse and thereby to criticize these equations, they accepted them on the basis of authority.

The J distribution breaks this deadlock by implying that notions of stability are best applied at the community level and not at the level of populations. Species are constrained only at the high abundance end by logistic influences embedded in the myriad interactions among species within a community, between communities, and between species and the physical environment. No species can have more biomass than will be found in a whole community of its fellows.

Under the Stochastic Species Hypothesis, random fluctuations in a population will leave a characteristic signature on plots of fluctuations in abundance, a binomial/normal shape as a manifestation of the central limit theorem (Hays and Winkler 1971). Section 7.2.2 will illustrate the appearance of a binomial/normal shape for time series of abundances for a variety of animal species, such data for other kingdoms being rather scarce.

1.6.1 THE RACETRACK ANALOGY

As we have already seen, a population may be viewed as vibrating in a stochastic manner, that is to say, both increasing and decreasing. The corresponding point on the abundance axis moves right or left as it increases or decreases. Viewed in time-lapse mode, the point would jitter back and forth, often drifting, and all in an uncertain and unpredictable manner. Large populations vibrate much more quickly than small ones because there are more individuals to die per unit time and correspondingly more to be born. Such events visit small populations, especially much smaller populations, much less frequently. Consequently, they vibrate rather slowly.

Imagine now a peculiar racetrack. Its length is divided into speed zones, all of equal length. For example, there is a 30 km/hr zone, a 20 km/hr zone, a 10 km/hr zone, a 100 km/hr zone, and so on, all in no particular order. From time to time, a car leaves the starting line at a random moment. Whichever speed zone it enters as it goes around the track, it must travel at that speed. After a considerable number of cars have entered this strange race, an aerial view of the track would reveal a strange distribution of race cars; the high-speed zones would have relatively few cars, since a car spends much less time in this zone. On the other hand, the low-speed zones tend to be rather crowded, since cars take much longer to get through them. In fact, it is easy to prove that the number of cars per zone, although fluctuating, would follow a hyperbolic distribution.

The racetrack analogy may strike some readers as far-fetched, owing to the presence of speed limits). However, if we let them "vibrate," going forward and backward by a uniform amount, but at random, the same phenomenon would be observed.

The cars in the 100 km/hr zone will now vibrate at twice the frequency of the cars in the 50 km/hr zone, tending to "escape" into an adjacent section of the racetrack about twice as quickly and hence lingering for a shorter time. The result would be, on average, only half as many cars in the 100 km/hr zone than in the 50 km/hr zone. This sketch may throw some light on the proof of the hyperbolic equilibrium theorem in Section A.1.

2 The J Distribution and Its Variations

In Chapter 1, the J distribution appeared only in the form of its probability density function (pdf). In this chapter, it appears as a distribution populated by a number R of species. The distribution comes in a continuous version and in a corresponding discrete version that translates the continuous form into a prediction of the number of species to appear at each possible abundance. Carrying this process one step further, an equivalent formulation called the *canonical sequence* specifies average positions for each species as a function of rank order.

In this chapter and the ones that follow, I will frequently use either form of the parameter delta, depending on the context. Readers need merely keep in mind that $\Delta = 1/\delta$ by definition.

2.1 PROPERTIES OF THE PROBABILITY DENSITY FUNCTION

Recall the density function in Equation 1.1 of Section 1.1:

$$f(x) = c\left(\frac{1}{x+\varepsilon} - \delta\right); \quad 0 \le x \le \Delta - \varepsilon$$
$$= 0; \qquad\qquad\quad x \ge \Delta - \varepsilon$$

It may cause some consternation to have a density function with a finite domain. Can there be no abundances greater than $\Delta - \varepsilon$? This question will be answered in Section 2.2.

The normalizing constant c is a function of the two parameters, ε and Δ, and does not constitute a new parameter, being merely a notational simplification:

$$c = \frac{1}{\left(\ln\left(\dfrac{\Delta}{\varepsilon}\right) - 1 + \dfrac{\varepsilon}{\Delta}\right)}$$

The constant c ensures that the integral of f encloses a unit area, as must all density functions. In what follows, the simplified (closely approximate) form of the formula is adequate to the task of playing the role of a normalizing constant.

$$c \approx \frac{1}{\left(\ln\left(\dfrac{\Delta}{\varepsilon} \right) - 1 \right)}$$ (2.1)

In most cases, the (absolute) approximation error would be less than 0.01.

2.1.1 THE CAPACITY CONSTANT

The constant c yields a closely related constant C, the *capacity* of the community, defined as the inverse of c. Inverting Equation 2.1 results in the following expression:

$$C = \ln\left(\frac{\Delta}{\varepsilon} \right) - 1.$$

Clearly, C is inversely proportional to the (log of) the twin parameters. It gets larger when the parameter Δ gets larger or the parameter ε gets smaller. The capacity C is smaller in the opposite situation. The difference between the two possibilities is illustrated in Figure 2.1a.

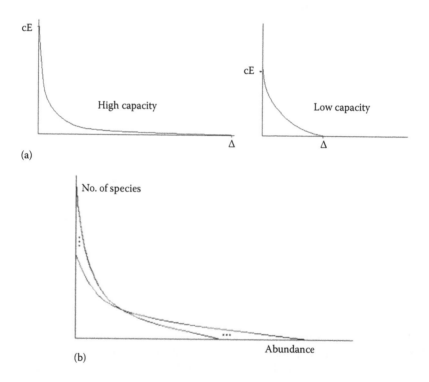

FIGURE 2.1   (a) Examples of high (left) and low (right) capacities. (b) Different shapes may share the same c value.

If, on the other hand, the capacity is kept constant, the parameters ε and δ may nevertheless vary, as long as the product δ/ε remains constant, as in Figure 2.1b.

For each fixed value of c, the range of density functions produced by varying ε and δ in this manner all belong to the same capacity class. The concept of capacity will be revisited in Section 2.2.2, where it plays a role in the spacing between consecutive average abundances; the greater the capacity C, the greater is the gap between the abundance of the kth species and that of the k + 1st. In distributions having a larger capacity, there is more "room," abundance wise, for additional species. Apart from appearing in Section 2.2.2, the concept plays a relatively minor role in the rest of this book. However, it strikes one as having promise for a part to play in later developments of the hyperbolic theory.

2.1.2 THE MEAN AND VARIANCE OF THE J DISTRIBUTION

According to the standard definition of the mean as the first moment of an (arbitrary) probability density function g, we have,

$$\mu = c \int_0^{\Delta-\varepsilon} \left(\frac{x}{x+\varepsilon} - \delta x \right) dx,$$

where $g = c(1/(x + \varepsilon) - \delta)$. The integral is evaluated in Appendix A.2.3 using the method of substitution. The resulting formula is

$$\mu = c\left(\Delta/2 - \varepsilon \ln(\Delta/\varepsilon) + \varepsilon^2/2\Delta\right). \tag{2.2}$$

Expanding the constant c in terms of Δ and ε does not result in a simpler expression, so it will be left in this form. In order to find the number N of individuals represented by a given theoretical distribution, simply multiply the value for μ just obtained by R, the richness of the sample.

The variance of the J distribution is based on the second moment about the mean:

$$V = c \int_0^{\Delta-\varepsilon} (x-\mu)^2 \left(\frac{1}{x+\varepsilon} - \delta x \right) dx.$$

The resulting formula, as derived in Appendix A.2.4, is not especially elegant:

$$V = c\left[(\Delta^2/2 - 2\Delta\varepsilon + (2.5)\varepsilon^2 - \delta(\Delta-\varepsilon)^3/3) \right] + \varepsilon^2 - \mu^2.$$

The variance of a distribution function F(x) = Rf(x) is obtained by dividing this expression through by R − 1. Owing to its noncentral nature and long tail, the J distribution has high variance. This is true of all previous proposals for species-abundance distributions, as well. Given the J-shape, it may be wondered if variance will be of much use in the context of the hyperbolic theory.

2.2   THE J DISTRIBUTION: SPECIES AND INDIVIDUALS

The distribution F for the J density function f is obtained by multiplying it by R, the number of species in a sample—or in a community, depending on the application.

$$F(x) = Rf(x)$$

Because the domain of the function is finite, one must interpret F carefully. In general, the integral of F over a given interval of abundances yields the expected number of species having abundances in that interval. Statistical fluctuations guarantee that some species in both samples and communities will usually *not* be found in their canonical positions (see Section 2.2.2). This includes the last canonical position, the logistic limit Δ, which bounds the average maximum abundance. Operationally, as in curve-fitting procedures, abundances beyond Δ are treated as if the J distribution had value 0 there.

The relationship between the value of the parameter ε and the number F_1 of species in the minimum abundance category, as expressed in the continuous versions of the J distribution, hints at the somewhat more subtle nature of the parameter ε:

$$F_1 = Rc(1/\varepsilon - \delta) \quad \text{or} \quad Rc(E - \delta). \tag{2.3}$$

Of course, c is already a function of epsilon. If the right-hand side of this expression is expanded in terms of the constant c, a mixed log/linear equation results that has no closed-form solution. The equation can be solved by numerical methods on a computer using the largest abundance as an estimate for Δ (or $\delta = 1/\Delta$) and the number of lowest abundance species as an estimate for F_1. This approach will yield a good approximation, in general, for the corresponding value of ε (see Appendix A.2.8).

2.2.1   THE DISCRETE J DISTRIBUTION

The discrete version of the J distribution is the main tool in fitting the J distribution to field data. Only when one has estimates in hand for the expected number of species in each category of abundance can goodness-of-fit methods be applied to a field data histogram. Early in the development of the J distribution, the discrete version played an important role by making the need for a second parameter (ε) obvious, as explained in Appendix A.1.2. The form of that parameter was dictated by the mathematics of the situation, ending in a symmetrical form wherein both parameters amounted to translations or displacements of the standard hyperbola.

To produce a discrete version of the J distribution, one first selects appropriate abundance categories, depending on the data at hand. Most commonly, the field worker will use categories 1, 2, 3, etc., representing simple counts: how many species appeared once in the sample? How many twice? And so on. But observations may also be grouped, so that, for example, the categories might be 1 to 3 for the first category, 4 to 6 for the next, and so on. Another common abundance format is density. What is the average number of Robber Flies observed per hectare? One species might have a density of 3.7, while another has density 42.6. Densities are normally grouped. Thus, the first (lowest) abundance category might run from 0 to 1.5, the next from 1.5 to 3.0, etc. The width of each abundance category will be called the *span* and written either as an integer or decimal number, depending on the context. Examples of such groupings will be found in the illustration in the Introduction.

In a general approach, the span will be a real number, a, implying the half-open intervals (0, a], (a, 2a], (2a, 3a], and so on. With this notation, one may produce a discrete distribution function by integrating over each abundance interval (category), as follows:

$$F(ka) = \int_{(k-1)a}^{ka} F(x)\,dx = Rc\left[\ln\left(\frac{ka+\varepsilon}{(k+1)a+\varepsilon}\right) - \delta a\right]; \quad k = 1,2,3,\dots.$$

Given values for R, c, k, and a, the function on the right is readily computed using a hand calculator, although a purpose-written program to achieve the same end is considerably more convenient. (See the program HGen in Appendix A.3.4.) When this is done for each category, a histogram like the one in Figure 2.2 will appear. The first category (k = 1) has the greatest number of species. The numbers decline as k increases, abruptly at first, more gradually later, to zero. In this idealized situation, the number of species in a given category might well be fractional. The number represents the expected or average number of species in this category over all instances of samples that this particular theoretical curve would fit the best. Chapter 4 explains how the discrete form is applied to field data.

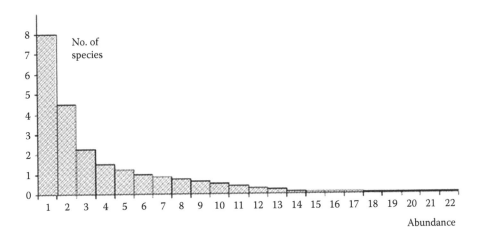

FIGURE 2.2   A discrete version of the J distribution.

A second form of discrete distribution may be derived independently by replacing the continuous variable x by the discrete variable k. The formula that emerges has almost the same values as yielded by its continuous relative:

$$F(k) = Rc(1/(k+\varepsilon) - \delta).$$

2.2.2 THE CANONICAL SEQUENCE

Figure 2.3 shows a set of idealized positions for abundances in a theoretical distribution (the thin, flat superimposed curve) that represents a hypothetical community. Each species is represented by a bar of unit height at a position given by formula (v), about to be derived.

The position of each bar approximates the actual expected positions, as shown in the plot of Figure 2.4.

Those who worry that the hyperbolic shape in samples indicates a large number of species about to be extirpated may take some comfort from these figures. Not only are species that appear only once in a sample likely to be more abundant in the source community than in the sample, but the lowest abundance there is not necessarily 1, as either figure makes clear. Over time, species come and go from communities, now appearing by immigration or now disappearing by emigration or extirpation. By the same token, at the scale of larger landscapes, species may appear by speciation or disappear by extinction. Chapter 9 examines the role of the J distribution in evolution.

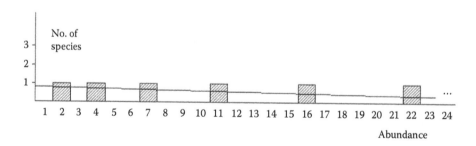

FIGURE 2.3   An idealized J distribution for a community.

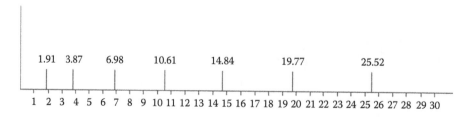

FIGURE 2.4   Canonical abundances of species in a small community.

The discrete version of the J distribution given in Section 2.2.1 describes the expected numbers of species per abundance category. Another description of the distribution lists the expected abundances themselves. The canonical abundance a_k of the kth species (in abundance rank) is readily found from the integral equation,

$$Rc \int_0^{a_k} \left(\frac{1}{x + \varepsilon} - \delta \right) dx = k. \qquad (2.4)$$

Appendix A.2.4 contains the full derivation of the resulting formula,

$$a_k = \varepsilon(\exp(kT + \delta a_k) - 1), \qquad (2.5)$$

where a_k represents the abundance of the kth species and $T = C/R$. Recall that C is the capacity of the distribution, defined as the inverse of the constant c. The maximum abundance is given by a_R. Because Equation 2.5 is not directly solvable, one may use iterative methods to zero in on a solution. Alternatively, resort may be had to the following approximation which applies most accurately to lower abundances:

$$a_k = \varepsilon(\exp(kT) - 1). \qquad (2.6)$$

Because of its simplicity and ease of application, Equation 2.6 may be preferred for some applications.

Worked example: Let $J(2.00, 150.0) \times 50$ be the J distribution corresponding to a sample. Then $c = 0.3003$ and $T = 1/15.015 = 0.06660$, so that

$$a_k = 2.0(e^{0.0666k} - 1.0).$$

Table 2.1 displays the first five canonical abundances corresponding to this distribution, as well as the last four. As the table shows, more than one species can have an expected abundance that is less than 1. At the discrete distribution level, the number of such species merely reflects the expected number of species to show up once in the sample. In this case, the number is 6. As it happens, some four species would be expected to show up with abundance 2 in the sample, and so on. At some point, we

---

TABLE 2.1

Canonical Abundances for a Sample of 50 Species

k	1	2	3	4	5	...	47	48	49	50
Abundance	0.138	0.285	0.442	0.611	0.790		63.4	72.0	83.4	99.8

must switch from Equation 2.6 to the exact or "general" expression Equation 2.5. The last four canonical positions were calculated by this method.

Finally, it should be noted that the canonical sequence formulation is simply another way of presenting the J distribution, being mathematically equivalent to it. The approximate formula used here is essentially a geometric progression, similar to the proposal of May (1975) discussed earlier in Section 1.2. I must also insert the following caveat: defining the canonical abundance of the kth species (in order) as occurring when the integral of Equation 2.6 has the value k may not be quite correct. The resulting mean abundance of the kth species may be slightly lower than the value given by Equation 2.5 but subsequent developments based on the integral values k chosen in Equation 2.5 will not be altered enough to invalidate any of the principal conclusions arrived at in this chapter.

The canonical sequence for a given J distribution, $J[\varepsilon, \Delta] \times R$ illuminates the use of the word "capacity" for the constant C. Recall that

$$C = \ln(\Delta/\varepsilon) - 1.$$

According to Equation 2.5, for the canonical abundances, we may replace the factor T by 1/Rc to obtain the formula

$$a_k = \varepsilon(\exp(kC/R) - 1.0),$$

which readily yields an expression for the space between consecutive abundances, as follows:

$$a_{k+1} - a_k = \varepsilon(\exp((k+1)C/R) - 1.0) - \varepsilon(\exp(kC/R) - 1.0)$$

$$= \varepsilon(\exp(kC/R))(\exp(C/R) - 1).$$

With ε and R both fixed, the expression is obviously larger when C is larger and, conversely, smaller when C is smaller. In other words, there is more "room" between successive abundances in high capacity communities. This concept should not be confused with the capacity of the community's environment to support its many individuals.

2.2.3 AN IMPLICIT FORMULA FOR RANK ABUNDANCE

Much of the prior work on species abundances has been framed in the context of the rank abundance diagram, a method of representing sample data that holds the same information as the standard representation, but in a rather different manner (see Section 4.2). One arranges all the species in one's sample in order of decreasing abundance (the rank order), then plots their logarithms as vertical bars. This section notes in passing that the rank abundance formula is implicit in the expression from

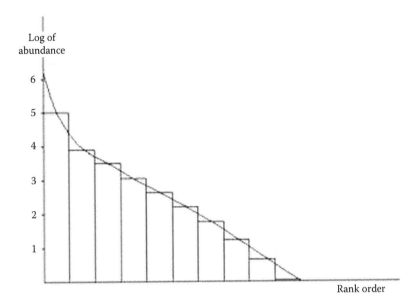

FIGURE 2.5   A simplified example of a hyperbolic rank abundance diagram.

which the canonical sequence is derived in Appendix A.2.4. The approximate version of this formula is solvable as

$$a(k) = \varepsilon(\exp(kT) - 1), \text{ where } T = 1/RC$$

where k is the index of the species in ascending order of abundance.

The order of the variable k must be reversed since the diagram begins not with the smallest abundances, but the largest. If A(k) represents the resulting function, then the following formula enables one to calculate bar heights for each rank abundance:

$$A(k) = a(R - k + 1). \tag{2.7}$$

The resulting curve can at least be illustrated through a worked example, as shown in Figure 2.5. Using the earlier example of J[2.0, 150.0] × 50, Equation 2.7 results in the plot where, for the sake of simplicity, only multiples of 5 are plotted as values for k. Although it is based on a rather small example, the resulting rank abundance diagram has the sinusoidal shape that is typical of such diagrams when based on actual samples.

It is not clear what role the rank abundance diagram has to play in the hyperbolic theory, but this example serves to illustrate how more than one abundance distribution can produce a rank abundance distribution with a sigmoidal shape.

2.2.4   EFFECT OF A LOG TRANSFORMATION ON THE J DISTRIBUTION

A variant of the J distribution that I shall likely never use is included here partly for the sake of completeness and partly because it results in what is known generally in the literature of population biology as the lognormal distribution (Magurran 1988). When a field sample of the kind being studied in this monograph is subjected to a logarithmic transformation, it frequently resembles a truncated normal distribution. In such a transformation, abundance categories are grouped, with abundance 1 enjoying a category of its own, abundances 2 and 3 occupying the next category, abundances 4 to 7 occupying the next category, and so on, with a doubling at each step. Preston (1948) may have adopted this representation of abundances in order to avoid the off-page outliers that are inevitable with the standard representation. A heavy price is paid for such convenience in the form of the information that is lost when 2^k abundances are added together to produce a bar height for the kth category under this representation.

However, it may be asked what happens if one subjects the theoretical form of the J distribution to a logarithmic transformation of the abundance axis, as used in connection with the lognormal distribution (see Section 4.2). The answer came as a surprise to me (Dewdney 1998b). A truncated, unimodal curve emerges that in many cases will appear bell shaped, albeit seemingly truncated on the left. To put the answer on a solid footing, the principal tool will be the general integral of the J distribution over the arbitrary interval [a, b]: Here, R' will represent the area under the curve that yields the number of species having abundances in this interval.

$$R' = Rc \int_a^b (1/(x + \varepsilon) - \delta)$$

$$= Rc\left(\ln((b + \varepsilon)/(c + \varepsilon)) - \delta(b - a) \right)$$

Values for the integral may be calculated for the subintervals (0, 1], (1, 3], (3, 7], and so on, according to the scheme of Preston, but this time, applied to a continuous function. It would yield essentially the same result if I used a discrete version of the distribution; the integral is simply a lot less work, with only one calculation per octave. Figure 2.6 displays the result when the log transformation is applied to the distribution J[2.0, 20.0] × 50.

According to the hyperbolic theory, samples subjected to this treatment would resemble a perturbed version of the histogram of Figure 2.6. The species that go missing from a sample are "veiled" by a sigmoidal curve when viewed on standard axes (Dewdney 1998b) and not by the veil "line" of Preston (1948). Researchers, such as Hubbell (2001), who use this representation run the risk of mistaking the log-transformed J distribution for the lognormal. Indeed, Gaston (2005) has pointed out that the shape that results from logarithmic axis compression cannot be Gaussian (i.e., normal). The example in Figure 2.6 illustrates the dangers inherent in a treatment that destroys information. Additional information about how this understanding of sampling works is found at the end of Section 3.5.

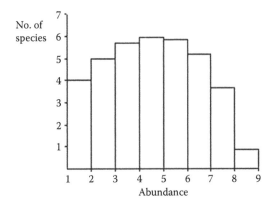

FIGURE 2.6   Log transform applied to abundance axis of the J distribution.

If one applies the same transformation to the distribution that most closely resembles the J distribution, namely, the log-series (Appendix A.2.7), one obtains a curve that does not resemble the lognormal but is triangular instead, with an initial peak, then descending asymptotically to the log-abundance axis.

3 Sampling in Practice and in Theory

A major focus of this book, the relationship between "communities" and "samples" in terms of abundances, must now be put on a more exact footing by defining both terms, both in theory and in practice. Since randomness plays an important role in the dynamics of communities, as well as in the sampling process, this concept must also be defined, at least empirically, as in the first section.

3.1 RANDOMNESS AND RANDOM NUMBERS

Most mathematical concepts applied to numbers are defined on the basis of the *presence* of certain properties. For example, a whole number is called *even* if it has 2 as a divisor; i.e., 2 is present among the factors of the number. Strictly speaking, a number is called *random* if it is obtained in the *absence* of any systematic procedure. Of course, no single integer can ever be "random" by itself. Randomness is always defined in relation to a sequence of numbers, with an implicit process of observing or producing members of that sequence repeatedly. Such sequences may arise deterministically or nondeterministically.

The only rigorous definition of randomness in relation to deterministic processes such as computer programs is based on the length of program that produces such a sequence. According to the only widely accepted definition of randomness (Chaitin 2001), a numerical sequence of length n is *random* if the minimum length of computer program that produces it increases as a function of n. In this definition, the sequence is potentially infinite. As (minimum-length) programs are found to generate longer and longer stretches of the sequence, it is found that the lengths of such programs grow in proportion to the lengths of the sequence. This definition has little practical value, as the implicit test (writing all possible computer programs that produce a sequence, then selecting the simplest) is far too complicated even to contemplate.

The following empirical definition of a random sequence makes no distinction between deterministic and nondeterministic sources but is certainly suitable for most scientific applications. A sequence is *effectively k-random* if all subsequences of length k have an equal probability of appearing in the (larger) sequence generated by the procedure. The following little observation tells us that effective randomness is a hierarchical property. The proof will be found in Appendix A.2.5: If a procedure is k-random, then it is also (k − 1)-random, for k > 1. This result paves the way for a full definition of effective randomness: A procedure is *effectively random* if it is effectively k-random for all values of k that apply. Any suitably long truly random sequence must be effectively random, but the converse is not necessarily true.

3.1.1 GENERATING RANDOM NUMBERS

Perhaps the most reliable process for generating (truly) random numbers must be inherently nondeterministic, exploiting a physical process such as quantum fluctuations of a vacuum field (Symul et al. 2011). However, such exotic sources are hardly necessary in the present context.

In practice, the "random numbers" generated by a computer are produced by a program known as a *pseudorandom generator*, the sequences so produced being called *pseudorandom*. A traditional method of generating pseudorandom numbers is the *linear congruential generator*:

$$X_{n+1} = (aX_n + c) \bmod m.$$

Starting with an initial or "seed" value X_0 for X, the program simply reiterates this basic equation, using the output of one iteration as input for the next one. The number m is called the *modulus*. Whatever the value that $(aX_n + c)$ might have, it is divided by m and the remainder taken as X_{n+1}. The constants *a* and *c* are called the *multiplier* and the *increment*, respectively. By choosing values for these constants appropriately, the stream of numbers so produced look convincingly random, even though technically speaking, they are not. Linear congruential generators are not much used these days, but the pseudorandom number generators of choice are not much more complicated. The random number generators in general use are effectively k-random for at least low values of k. They are presumably suitable for the experiments described in this book since they are suitable for commercial simulation (an industry in its own right) as well as most scientific simulation of which the author is aware.

One of the most influential sources of randomness in nature is the weather. Given that weather systems tend to be chaotic, weather variables tend to vary unpredictably over time in any given area or region. Although weather systems appear to have a Lorenz attractor (Lorenz 1963) at their dynamical core, much simpler chaotic systems yield the same random effects. For example, one can even use the logistic map (May 1976, Verhulst 1838) to generate pseudorandom numbers within the interval [0, 1]. Larger numbers may be derived from these by multiplying the outputs by a suitably large constant:

$$X_{n+1} = \lambda X_n (1 - X_n). \tag{3.1}$$

Verhulst and May both studied the logistic map as a potential source of insight into the behavior of populations. Equation 3.1 governs the direction and extent of population changes over time: If the populations N_1 and N_2 of two organisms sum to a fixed number N (the logistic limit), the populations may be expressed as ratios $X = N_1/N$ and $Y = N_2/N$. Equation 3.1 uses subscripts to indicate successive values of the variable X. It uses, as well, the obvious relation, $Y = 1 - X$.

In order for the equation always to produce values that lie in the interval [0, 1], we require that

$$0 \leq \lambda \leq 4$$

since the right-hand side of Equation 3.1 takes its maximum value, 0.25λ, when X = 0.5. The parameter λ could be called the *fecundity factor*, as it strongly influences the rate at which either population may grow. At low values of λ, the variable X quickly converges to a specific number and remains there. At higher values of λ, the variable X alternates between two fixed numbers, then four, then eight, and so on, as λ is increased. The period-doubling behavior of the logistic map continues right up to a value of approximately 3.57 for λ, where chaos sets in. The variable X bounces around inside the interval [0, 1] with no seeming rhyme or reason. This feature of the equation is responsible for its popularity as an early population model in ecology. Its appeal lay in its seeming realism, behaving just as unpredictably as real populations.

A computer program that embodies the logistic map was used to generate 200 numbers in the interval [9, 1], of which a randomly selected subsequence of 10 consecutive numbers are listed here as an example:

0.7681   0.6768   0.8310   0.5335   0.9457   0.1951   0.5966

0.9156   0.2971   0.7935…

The numbers were produced by the logistic map with the parameter λ set to 3.8. In order to illustrate the presence of effective randomness in such numbers in a binary setting, each number has been replaced by the parity (even or odd) of the sum of its digits: 0 1 0 0 1 0 0 1 1 0…

Performing a spectral analysis of the resulting bits, the sequence produced 51 0-bits and 49 1-bits, making it fairly 1-random, while the spectrum of consecutive pairs produced the frequencies shown in Table 3.1, again well within what might be expected from a 2-random source. The number of observations in each of the four categories should follow the uniform distribution, converging to relative equality as more and more effectively random numbers are generated. Before going on to examine higher orders of effective randomness from such a source, the sequence of 200 "chaotic" numbers would have to be doubled, then doubled again, as the analysis proceeded. Most tests for randomness, from the poker test to the runs test (Law and Kelton 1999), are essentially specialized versions of the spectral test.

TABLE 3.1

Frequencies of Two-Bit Subsequences in the Random Sequence

Subsequence	Frequency
00	14
01	13
10	10
11	13

3.2   COMMUNITIES AND SAMPLES

A *natural community* must be defined somewhat loosely to accommodate the many kinds of "communities" studied by field biologists. Briefly, it consists of all species belonging to a group G and living within a specified area (or volume) of a natural environment over a specified period of time. The loosest part of this definition is clearly the group G. Typically, the grouping G is based on taxonomy, as in all species of Basidiomycota growing or present in a north temperate woodlot of specified dimensions or limits. G may also be trophically based—or partially so—as in all seed-eating arthropods in a tropical grassland, again of defined extent. In all cases, there is a geographic habitat restriction. Of course, each species in such a natural community is represented by the individuals that belong to it, whether many or few in number.

A *theoretical community* is much simpler. It consists of a finite collection of (abstract) species, each of which consists of a given number of individuals. The species are distinguished by an index number, perhaps, but the individuals are normally represented only by a count. A habitat is of course assumed.

From either kind of community, one may take a sample. If a community consists of N individuals (of various species), then a sample of size n consists of a random drawing (or observation or collection) of n individuals from the larger population of N individuals. The sample is unbiased if, when repeated many times, it shows no overall tendency to favor one species disproportionately to its actual abundance. The sample is *with replacement* if individuals are not actually removed from the community; otherwise, the sample is said to be *without replacement*. Here, we must also introduce the concept of *intensity*, namely, the ratio $r = n/N$. It will be shown presently how intensity may be estimated in the context of a field study. This definition of sample may be applied to either natural communities or theoretical ones. The connection between the two definitions just given is that a natural community, as defined operationally, automatically implies the set structure of the theoretical definition. The structure is largely invisible, of course, and only to be discerned, dimly enough, through samples taken of it.

3.3   THE SAMPLING PROCESS

If a sample is "random" enough, the biologist may be fairly confident that a species that is relatively common in the sample is also relatively common in the community. By the same token, a species that occurs just once in a sample will tend to belong to a relatively small population in the community. In most cases, many of the lowest-abundance species in the community will not show up in a sample at all. To ensure that the sample follows the community in the foregoing statistical sense, the person sampling must of course eliminate bias, taking compensatory steps to eliminate it from the sample. For example, if one is sampling a community of butterflies, some species may be drawn preferentially to a particular species of plant. Sampling patches of such plants exclusively will cause the species in question to be overrepresented in the sample. Sweeping a large, defined area with the greatest variety of plant types will certainly help to eliminate that particular source of bias—as long as the sweeping trails are randomly selected (and recorded for subsequent resampling, if necessary). Section 3.3.1 makes clear how both sampling with replacement and sampling without are used in taking samples, depending on the target organisms.

The intensity r of one's sample is a key factor in understanding the sampling process. As will be shown in Chapter 5, no richness estimation technique can work reliably in the absence of knowledge of r. If one does not know the intensity of one's sample, no method exists—or can exist—that will estimate with (statistical) accuracy the number of species in the community being studied. Indeed, the accuracy of any such method will depend critically on the accuracy of one's estimate of r. In Section 3.3.2, some simple methods are given for arriving at accurate estimates of intensity, at least for relatively simple kinds of community.

It was common in the mid-to-late twentieth century to assess the "biodiversity" of a community. Section 3.3.3 indicates why the project fell on hard times. Too many kinds of biodiversity were defined, and they all meant different things, leaving none as a principal tool except the simplest of all, namely, richness R, the number of species in the community being sampled.

3.3.1   THE VARIETY OF SAMPLING ACTIVITY

The most common forms of sampling with replacement involve plants and some animals such as birds or fish. A botanist carrying out a survey of oldfield plants, for example, may lay out several quadrats at random and record every plant contained within each quadrat without picking any of them. (Not removing a plant at all is equivalent to sampling with replacement, although an occasional voucher sample may be taken.) A point-count of birds is carried out by the zoologist or bird expert walking a specified distance, then stopping for a specified period of time and noting all birds calling or visible within a given distance. Of course, in this case, there is no actual collecting done, but the technique amounts to the same thing. The birds are still there when the biologist leaves the area. A somewhat better example is the method of sampling bats by catching them in a mist net when they are flying. A specimen can be extracted from the net, identified, recorded, and then released. There is always the problem of overcounting with these techniques, although for bats and some other animal species, the problem can be solved by marking the specimen with a temporary stigma before releasing it. Similar remarks apply to some forms of sampling fish. Most fish-trapping techniques, from seine nets to kick-samples to minnow traps to drag nets, can be deployed in this manner. Electrofishing, however, may injure the fish brought to the surface (stunned) by powerful electric currents. Small mammals such as moles, shrews, voles, (nonjumping) mice, and small mustelids may be captured in pitfall traps, again to be counted and released. In all the foregoing cases involving vertebrates, the group being sampled is usually well known and an expert field biologist can make an identification rapidly enough to release the animal before any harm is done.

Fungi occupy a special place in this summary of techniques because, at least with macrofungi, only the fruiting body is normally collected, leaving the organism with its mycelial network more or less intact. Microfungi are typically part of soil samples or samples of other substrates. They may be identified by microscopic examination in situ or in cultures or identified through DNA analysis of substrate samples.

Sampling without replacement is most common in arthropod surveys. No other group of organisms has more collecting techniques applied to it and almost all involve sampling without replacement, "sacrificing" the animal, if you will. Arthropods may be captured in light traps, pitfall traps, malaise traps, Berlese funnels, emergence traps, aspiration bottles, and pan traps and by sweeping, fogging, hand-collecting, beating, observing, and so on. We note here that different collecting techniques carry a natural emphasis on different habitats. Night-flying insects may be drawn to a light trap, but insects that are not flying at the time will not appear there. Indeed, moths, for example, may be drawn preferentially to a light trap, some strongly, some only weakly. Such preferences undoubtedly introduce bias into the sample so taken. Ground beetles (Staphylinid, Carabid, etc.) are natural victims of pan traps and pitfall traps, but other beetle families may be collected by sweeping bushes with a net. One should add photography to this list of collecting techniques, an increasingly popular sampling technique and obviously "with replacement."

For the smallest organisms, namely, protists and bacteria, a sample of soil or water is collected and transported back to the laboratory for examination. If not killed by staining or other chemical treatments, the organisms may be flushed into the sink once they have been examined, but they are never "replaced."

When populations are large and samples are relatively small (as in the foregoing paragraph), it makes little difference to the final assessment whether the sample was taken with or without replacement. Removing a few individuals at random from a population of millions has no discernible effect on the balance of probabilities for subsequent observations.

3.3.2   ESTIMATING SAMPLE INTENSITY

Readers will recall that the sample intensity r is simply the ratio of individuals in the sample to those in the community being sampled. In the case of plant surveys, r is easily estimated. Assuming that the field biologist has a defined polygon within which a survey will be taken, it is relatively easy to calculate r. For example, if the biologist records every plant in each of 10 randomly placed 1-meter square plots within the polygon, the intensity will be

$$r = 10/A,$$

where A is the area of the polygon in square meters. Mycologists collecting samples have roughly the same advantage as botanists when it comes to calculating r.

In the animal kingdom, the sampling process is bedeviled by the tendency of subjects to wander off or fly away, to be counted more than once or not at all. However, even here, some groups are easier than others. For example, a north temperate waste field in August might present a sea of wildflowers such as goldenrod, aster, wild carrot, etc., with clumps or patches of species mixed uniform-randomly throughout the field. The flowers are attended by several orders of insect, especially Hymenoptera and Diptera. The fact that bees, wasps, and ants are constantly moving from plant to plant may not matter, provided that the pattern of movement prevails over the whole field. Carrying out an intensive count within a number of randomly placed quadrats may give one a good estimate of the density d (individuals per square meter). This would lead to a reasonable estimate of the total size N of the hymenopteran meadow community, and the intensity of the sample will be

$$r = n/N,$$

where n is the size of one's sample. Variations on this technique may be applicable to other groups, the areal consideration being one component and the estimate of total population being the other. In general, the technique calls for intensive sampling (census would be a better word) within a limited area or areas, then extrapolating the result to the area under study.

A final caveat must be added to any description of intensity, as defined earlier. Some writers refer to sampling effort, and this may be taken as equivalent to intensity, provided that the same amount of time, energy, and movement patterns of the sampler is always the same over the study in question. In other words, if I sample hymenopterans according to a fixed protocol but spend longer at each counting station for one sample as I do for another, the two samples are not comparable. The definition of intensity given here must involve uniform amount of time, energy, and movements by the sampler, otherwise, compensatory steps must be taken. A similar remark applies to the activity level of the species being sampled. Plants are no problem, but a change in weather can have a marked effect on the visible abundance of insects, for example.

3.3.3   HOW SAMPLES HAVE BEEN USED: CALCULATING BIODIVERSITY

In the biosurvey literature, one frequently finds authors who wish to go beyond the mere presentation of the abundance data they have painstakingly gleaned from nature. They wish to say something meaningful about the sample, usually in terms of its "biodiversity." But to do so, they face a bewildering choice of "biodiversities" to choose from.

Although the term "biodiversity" has been much used in recent years, it turns out to have no generally accepted definition. Instead, it has many. This illustrates once again the "confusion" that pervades theoretical ecology, as explained in Section 1.4. Although these concepts of biodiversity are defined for sample data, they are frequently interpreted as descriptions of the communities from which the samples came. In the following, we consider a community (or a sample of it) with m species having abundances given by the integers $a_1, a_2, a_3, \ldots, a_m$. Some of the following biodiversity measures (indices) use abundances in relative form $p_1, p_2, p_3, \ldots, p_m$, where $p_i = a_i/N$, N being the total number of individuals in the community (or sample). The maximum abundance *max* is defined as the largest integer in the set $\{p_1, p_2, p_3, \ldots, p_m\}$

Simpson Index	$B_1 = \sum p_i^2$	(Simpson 1949)
Shannon Index	$B_2 = -p_i \ln(p_i)$	(Shannon 1949)

Shannon-E* Index $B_3 = -\dfrac{B_2}{\ln(\text{max})}$ (Pielou 1969)

Brillouin Index $B_4 = \dfrac{\ln(N!) - \sum \ln(a_i!)}{N}$ (Pielou 1969)

Margelef Index $B_5 = \dfrac{R-1}{\ln(N)}$ (Clifford and Stephenson 1975)

Menhinick Index $B_6 = \dfrac{R}{\sqrt{N}}$ (Whittaker 1977)

Berger-Parker Index $B_7 = \dfrac{\text{max}}{N}$ (Berger and Parker 1970)

There are still other measures of biodiversity, but they add nothing to the illustration. The first three indices use relative abundances, but the last four do not.

Three artificial communities will serve to illustrate some of the different results that the foregoing measures yield when applied to them. Figure 3.1 displays three histograms, each of which has 12 species and 36 individuals. The first histogram, named C_1, albeit somewhat reduced for illustrative purposes, is typical of samples that commonly emerge from collections in the field. The second histogram, C_2, is univoltine; all species have the same abundance. In the third histogram, C_3, all species except one have abundance one.

Although community C_1 is most typical, in overall shape, of real communities, none of the first five indices give it their highest biodiversity assessment. Instead, as shown in Table 3.2, both Shannon indices and the Brillouin index give the highest score to what can only be described as a distinctly pathological community, C_2, where all species have the same abundance. Since these three indices regard such a shape as in some sense the ideal, their use has limited value. The Berger-Parker and Simpson indices award the palm to the extremal community C_3, again indicating a lack of contact with real data.

The two remaining indices do not take individual abundances into account and so give the same score to all three communities. Given that the intention behind the first five indices is to take not only the richness but also the shape of a distribution into account, it must be remarked that developing a single numerical measure that embraces both aspects of a distribution is a doubtful project. It would be better to use at least two measures as a joint descriptor of "biodiversity." Otherwise, let that term stand as a synonym for species richness. As though the failings and confusion created by so many indices were already apparent, most field biologists already follow the latter practice.

Gaston (1996) has summarized the problem of defining "biodiversity" as follows: "However, the abstract concept of biodiversity as the 'variety of life,' expressed across a range of hierarchical scales, cannot be encapsulated in a single variable. The complexity in this sense is irreducible, and the search for the all-embracing measure of biodiversity, however desirable it might seem, will be a fruitless one." Gaston reminds us that "evenness" and "equability" were two of the goals of early biodiversity measures. Hurlbert (1971) has expressed similar views.

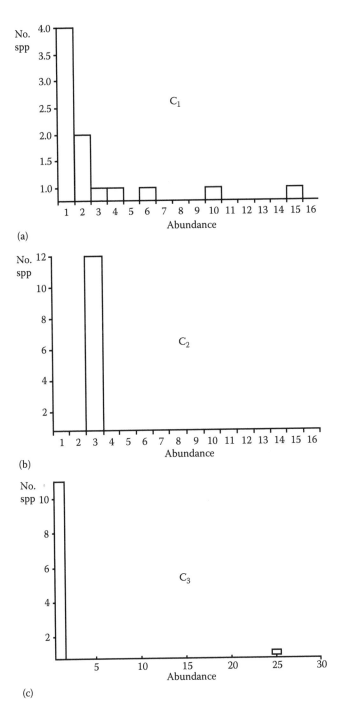

FIGURE 3.1 (a) A typical field histogram. (b) A univoltine distribution. (c) An extreme distribution.

TABLE 3.2

Biodiversity Indices versus Three Different Communities

Community	C_1	C_2	C_3
Simpson	0.116	0.084	**0.493**
Shannon	2.171	**2.485**	1.348
Shannon E	1.443	**29.81**	3.697
Brillouin	1.811	**2.062**	1.048
Berger-Parker	0.222	0.083	**0.694**
Margelef	3.070	3.070	3.070
Menhinich	2.000	2.000	2.000

Note: Bold numbers are maximum values achieved by a measure.

3.4 COMPUTER SIMULATION OF SAMPLING

It is hard to see how anyone carrying out research in theoretical ecology, at least in the area of species abundance distributions, can expect to understand the sampling process without the appropriate computational tools. The most important such tool is obviously one that simulates the sampling process itself. As a destroyer of hypotheses, it has no equal. The author has, on some occasions, seen a favorite hypothesis contradicted by simple computer experiments, whereupon the hypotheses were immediately abandoned.

Although most field biologists will not be simulating the sampling process, they should understand the relationship between "real" sampling and simulated sampling. Ideally, they are the same, but in practice, they may not be. The idealized sampling process described in this chapter is an exact match with standard statistics—to within the powers of pseudorandom number generators. In the context of the analysis of natural communities, the samples taken must be random enough to ensure that a meaningful sampling intensity figure can be calculated and applied. The tools described here are indispensable, however, for those who would understand sampling in an ecological context.

Suppose we have in hand a collection of N individuals from many species. This could be considered as a community, a portion thereof, or even the sample of a community. A random sample taken of the N individuals would be oblivious to which species the selected individuals might belong to. After the fact, one may construct a standard histogram, as described in the next chapter, and expect to see the community distribution reflected to some degree in the histogram.

In what follows, all samples will be with replacement, i.e., not removing any of the items of the sample from the original population (community) of N individuals. In other words, some individuals may be sampled more than once, in effect. This process is readily simulated.

3.4.1   A Sample Simulation Algorithm

Simulating a direct sampling process is easy by computer. It takes an abundance distribution as input, either theoretical in the form of an expression that generates expected frequencies or empirical in the form of a list or *biodiversity array*,

$$(a_1, a_2, a_3, \dots, a_R),$$

where a_i represents the abundance of the ith species and R is the number of species. The theoretical expression just mentioned can readily be converted into a biodiversity array, as used in the following algorithm. Random individuals are selected by generating a random number, then counting through all the species in the biodiversity array until that number is arrived at. Thus, if the number 27 is chosen at random and the first seven abundances in the array are

$$5, 9, 2, 8, 5, 5, 15, \dots,$$

one would count through the array entries by adding up the abundances along the way. When the last entry where the sum still falls short of 27 is reached, the next species is the one that individual 27 belongs to. Thus, we have

$$5 + 9 + 2 + 8 = 24 \text{ while } 5 + 9 + 2 + 8 + 5 = 29,$$

so the 27th individual belongs to the fifth species in the list. The algorithm that appears in the following uses three arrays, a source array of given species counts that will be the target of the sampling process, a sample array for holding the sample computed by the program, and a histogram array. The first two arrays will have the form of biodiversity arrays, while the histogram array will be indexed by counts (1, 2, 3, etc.) of species.

Parameters N and r will represent the total number of individuals in the source and the intensity of the sample, respectively.

Sampling algorithm:

1. Input biodiversity array, along with values for N and r.
2. Set all entries to 0 in the sample array and in the histogram array.
3. Let n be the integer {rN} (i.e., the greatest integer not exceeding rN)
4. Repeat the following steps n times:
   4.1. Choose a random number j from the interval [1, N].
   4.2. Calculate s as the species containing the jth individual.
   4.3. Increment the count for species s in the sample array.
5. For each species,
   5.1. Look up its abundance as calculated in step 4.
   5.2. Add 1 to the species count at that abundance in the histogram array.
6. Output the sample array or the histogram array, as desired.

The sample simulation algorithm just described involves sampling with replacement. To convert it into a program that samples without replacement, simply add the following instructions to Step 4:

4.4. Decrement the count for that species s in the source array.
4.5. Decrement N by unity.

One may obtain the "perturbation" of a distribution (see Section 8.1.1 for a formal definition) in this manner by setting r = 1. The algorithm just listed is the simplest possible. It may be implemented in any programming language one happens to be familiar with, but it is essential to follow the structure of calculations given here so as to ensure that individuals are sampled, and not species.

3.5 THE GENERAL THEORY OF SAMPLING

The most important thing to know about the sampling process from a theoretical point of view is that it expresses what mathematicians call a transformation or mapping, in this case the transformation of a community distribution, into an image set—the sample distribution. The transformation has the important property that it preserves abundance patterns in a special way. If the community abundances follow the distribution G in a statistical sense, then so do the abundances within the sample, albeit with different parameter values. Strangely, there was no general sampling theorem in the literature before 1998 (Dewdney 1998b). If the theorem had been published earlier, say in the 1930s or 1940s (when it could have been), the confusion resulting from multiple proposals could have been avoided.

The Pielou transformation is named in honor of Evelyn Pielou, a researcher in ecology who used a similar tool (Pielou 1977). The transformation maps the abundances in a community with the distribution F into an "expected" sample with distribution F', where

$$F'(k) = \int_0^\infty \left(\frac{e^{-rx}(rx)^k}{k!} \right) F(x)\,dx, \qquad (3.2)$$

the integral being taken from 0 to ∞. In Equation 3.2, r represents the intensity of the sample that links the two distributions. We note that the variable k denotes abundances in the community (F) but is treated as a constant in the sample (F'), whereas x plays the role of abundance variable in the community.

The transformation is based on a statistically exact formula, the hypergeometric distribution, which involves a ratio of factorials that are somewhat cumbersome to work with mathematically or to program, for that matter. However, the hypergeometric distribution is very closely approximated by the Poisson distribution (Hays and Winkler 1971), as it appears in Equation 3.2. One could argue that "exact ecology" should not involve approximations; however, reverting to the hypergeometric distribution is always possible and, if need be, one could determine the approximation error for specific cases. Thus, exactitude, although slightly compromised, is always within reach with some additional effort. The following example is intended to illustrate the action of the Pielou transform, in particular its function-preserving property.

From a purely mathematical point of view (sampling theory aside for the moment), one may ask what form the function F′ would take if we set $F(x) = x^n$, the simplest form of polynomial. Here, the indefinite form of the integral suffices to make the point.

$$F'(k) = \int \left(\frac{e^{-rx}(rx)^k}{k!} \right) x^n \, dx,$$

$$= r^{-n} \int \frac{e^{-rx}(rx)^{k+n}}{k!} \, dx,$$

$$= \frac{r^{-n}(k+n)!}{k!} \int \frac{e^{-rx}(rx)^k}{(k+n)!} \, dx.$$

Evaluated between its limits, the integral equals unity, leaving

$$F'(k) = \frac{r^{-n}(k+n)!}{k!}.$$

If the expression $(k + n)!$ is multiplied out, one obtains a new polynomial in k of degree n. For example, if $F(x) = x^2$, we have, with $n = 2$,

$$(k+n)!/k! = (k+2)(k+1)$$

so that

$$F'(k) = r^{-2}(k^2 + 3k + 2).$$

In this particular case, one may describe the effect of the transformation as follows. The function x^2 is an upward-opening parabola with its apex at the origin (0, 0), while F′(k) is an upward-opening parabola with its apex at the point (−1.5, −0.25). Moreover, the factor r^{-2} being less than unity has the effect of increasing the height of the parabola, just as sample histograms tend to be higher than those arising from the community. Summarizing the net effect, the Pielou transformation shifts the initial parabola 1.5 units to the left and heightens it in the process.

The transformation of a general polynomial of degree m may be regarded as the transformation of a sum of terms having the form x^n, for $n = 1, 2, ..., n$. The result is a polynomial of degree m that is compressed horizontally and compared to the general polynomial that we started with. One step remains in the chain of reasoning.

Let F be a theoretical species-abundance distribution that is continuous over its domain. The second part of the argument proceeds by invoking the Weierstrass

Uniform Approximation Theorem (Hobson 1950). Any continuous function F can be approximated over its domain to an arbitrary precision by a polynomial expression, according to the Weierstrass theorem. One may therefore replace any distribution F(x) by a certain polynomial P(x) having the property that

$$\left|F(x) - P(x)\right| < v$$

holds over the domain of F for an arbitrarily small quantity v. If we apply the Pielou transform to the function P, we obtain a polynomial P′ that can be made (by choosing v small enough) to approximate F′ to any desired degree of precision, as the following inequality makes clear:

$$\int \left(\frac{e^{-rx}(rx)^k}{k!}\right)\left|F(x) - P(x)\right| dx \le \int \left(\frac{e^{-rx}(rx)^k}{k!}\right) v\,dx = v. \qquad (3.3)$$

The integral on the right-hand side of Equation 3.3 is the area under the Poisson density function, namely, unity. The degree of approximation of F(x) by P(x) is therefore inherited by F′(x) and P′(x). The inheritance of shape by P′(x) from P(x) therefore implies an inheritance by F′(x) from F(x). In the limit, the approximation error is zero.

It now follows that for each value of k, the following two integrals can be made arbitrarily close by choosing P appropriately:

$$F'(k) = \int \left(\frac{e^{-rx}(rx)^k}{k!}\right)F(x)\,dx \quad \text{and} \quad P'(k) = \int \left(\frac{e^{-rx}(rx)^k}{k!}\right)P(x)\,dx.$$

The functions F and F′ therefore have the same general form, differing only in the values of their (common) parameters.

One application of the sampling theorem to date has been to imply that the "veil line" proposed by Preston (1948) is incorrect; the lognormal distribution, as portrayed on our standard species/abundance axes, would be mapped into another lognormal distribution and not a truncated lognormal distribution. Another, more important application occurs in Chapter 5, which describes a method for estimating the richness of a community based on its samples. The formula in the previous sampling theorem is used to calculate the sample of an estimated community distribution to compare with a sample. Parameter values for the community can then be changed in a direction that gives a better estimation, based on that comparison. In this manner, the parameters of the community distribution converge on values for which samples (via the formula) give the best possible match with the empirical data.

4 Compiling and Analyzing Field Data

No experienced field biologist needs to be told what to do in compiling a list of species and abundances for a specific area. Yet the same biologist may be unaware of what biologists in quite different areas of biodiversity assessment may be doing. In fact, there is a general pattern or template that can be applied to all sampling activities of this kind. The template gives me the opportunity of linking the general activity with the specific requirements of the J distribution.

When sampling in the field, one invariably keeps records, whether the organisms being sampled are identified on the spot, through photographic records, DNA samples, or taken back to the lab. Normally, the number of individuals within each species is counted. The raw counts may be used directly or converted to densities or percentages on an areal or volumetric basis. The J distribution is equally friendly to all forms of count or count surrogates. In what follows, I describe basic methodology for the arrangement and display of field data, as well as the two most widely used measures of fit for theoretical distributions versus field histograms. A method for estimating more accurately the true species overlap of two samples appears at the end of this chapter as an illustration of the potentially wider role to be played by the J distribution in the analysis of data.

4.1 HISTOGRAMS AND DISTRIBUTIONS

A histogram is a compilation of data into categories for the purpose of revealing a shape or trend that might not be obvious from examining the data as a mere list of numbers. In the case of abundances appearing in a sample, the most natural categories are ranges of values into which the abundances may be sorted. Here, in Table 4.1, for example, is a set of abundances of Microlepidoptera taken in a light trap in the Netherlands in 2006 (Jansen 2005). The abundances are displayed in Figure 4.1. In general, remaining abundances that do not appear in a table or histogram may be listed separately, as in Figure 4.2.

Some field data plot as a relatively shallow J-curve, with species gaps occurring even at low abundances. In such cases, it may be desirable to group the data by twos or threes.

Worked example: The histogram of abundances shown in Figure 4.2 represents numbers observed by Busby and Parmelee (1996) in their survey of herpetofauna in Kansas. It has the typically ragged shape of data in which fewer species, on average, inhabit the lower abundance categories.

When the categories are changed by grouping, the shape implicit in the data is more clearly seen. In Figure 4.3 we have grouped the abundances by fives. The remaining raggedness is normal, even in grouped categories, as seen here.

Often, a field biologist will report abundance data in terms of densities of individuals per unit area. Densities are no different from whole number counts in being subject

TABLE 4.1
Abundances of Microlepidoptera Taken at a Light Trap

Category	No. spp.	Category	No. spp.	Category	No. spp.
1	16	11	1	21	1
2	7	12	0	22	0
3	3	13	1	23	0
4	3	14	0	24	0
5	2	15	0	25	1
6	1	16	1	26	0
7	2	17	0	27	0
8	2	18	0	28	0
9	1	19	0	29	0
10	1	20	1	30	1

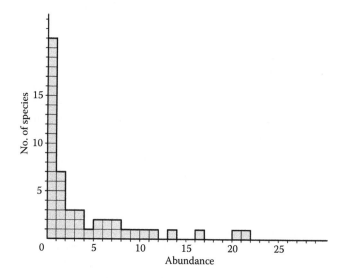

FIGURE 4.1   A species/abundance histogram of Microlepidoptera data.

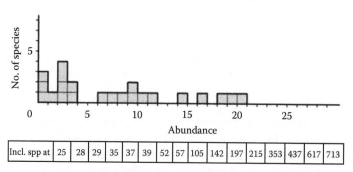

Incl. spp at	25	28	29	35	37	39	52	57	105	142	197	215	353	437	617	713

FIGURE 4.2   Field histogram of herpetofauna.

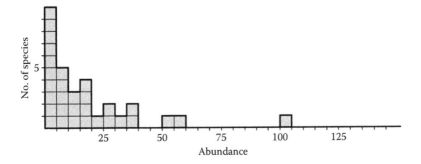

FIGURE 4.3 Histogram of Figure 4.2 after grouping operation.

to compilation into categories. The J distribution, being continuous, works just as well with these data as with whole number counts. In such a case, the abundance axis might well be labeled with decimal numbers such as 0.1, 0.2, 0.3, etc., for example.

4.2 OTHER REPRESENTATIONS: RANK ABUNDANCE

As explained in Chapter 3, the *rank abundance diagram* is obtained by placing abundances in decreasing order, taking their logarithms, then plotting them. For example, if one uses the abundance data from Table 4.1, takes their logarithms (to the base *e*), places them in rank order, and plots them, one obtains Figure 4.4.

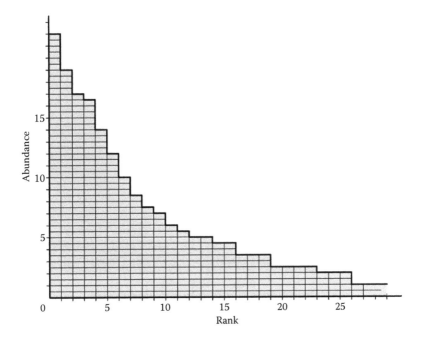

FIGURE 4.4 Rank abundance diagram for the data of Table 4.1.

Apart from the disadvantage of vertical scale compression, the rank abundance diagram is somewhat unwieldy mathematically speaking. The horizontal axis is not a metric axis, but an ordinal one, so metric operations such as grouping or scaling do not apply in any meaningful sense. Moreover, operations such as the insertion or deletion of a species, trivial to carry out in the standard diagram, involve shifting all the species on the right-hand side of the one deleted or inserted. In terms of the information stored in it, however, the rank abundance diagram is equivalent to the standard system—which can be reconstructed by reversing the process.

4.3   OTHER REPRESENTATIONS: LOGARITHMIC ABUNDANCE

The *logarithmic abundance diagram* does not contain information that is equivalent to the standard axis system. In this scheme, abundance categories are grouped into "octaves," the first octave consisting of the first or lowest abundance, the next octave consisting of the next two abundances, the octave following that one consisting of the next four abundances, and so on, with the kth octave consisting of the abundances 2^{k-1} to $2^k - 1$. Plotted in this manner, the Jansen data of Section 4.1 appears in Figure 4.5. Often, such data have the appearance of a truncated hump without the initial spike that appears here. I have labeled each category with the highest abundance in its group. If I used a base higher than 2, the hump may well appear.

The alleged charm of the lognormal distribution arises from the bell-shaped (i.e., unimodal) curve, usually truncated on the left, that emerges when species abundance

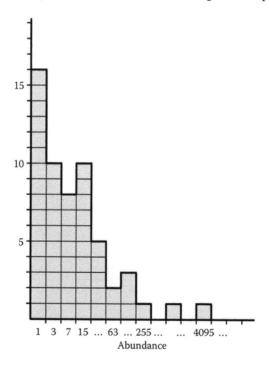

FIGURE 4.5   Histogram of data in Table 4.1 grouped by "octaves."

data are plotted by octaves. Sometimes (as previously), the shape does not emerge with any great clarity. As explained at the end of Chapter 2, however, if one subjects the J distribution to a logarithmic transformation, a truncated unimodal curve emerges (Dewdney 1998b). Moreover, the bell-shaped curves that appear under these circumstances are completely indistinguishable, at least by visual inspection, from those that arise from the lognormal distribution. In other words, the appearance of a truncated bell-shaped curve in this context does *not* permit one to conclude that the data subjected to the transformation follow the lognormal distribution, as the J distribution might be regarded as the better interpretation.

Apart from the slight advantage of being able to represent higher abundances using either the log-transformed representation or the rank abundance diagram, there is no particular advantage in using either distribution. All field data and theoretical curves are well represented by the standard axis system. As I have mentioned before, it has the additional advantage of being the conceptual theater, in which we may visualize stochastic vibrations directly, as it were.

4.4   ESTIMATING PARAMETERS OF THE J DISTRIBUTION

When the field biologist has taken one or more samples of the community under study, the first step in analyzing the data is to plot the species abundance histogram as described in the previous section. The next step, although not necessary for all purposes, is to find the best fit of the J distribution to the data as plotted. Such a procedure is essential in applying one of the richness estimation techniques to be described in the next chapter.

To find the best fit of field data to the J distribution, the easiest measure to use is the chi-square goodness-of-fit test. For an exact best fit, there is no alternative to testing more than one combination of values for the parameters ε and Δ. A method for finding an optimum fit to the J distribution is outlined in the next section. A reasonably good (suboptimal) fit may be obtained indirectly from the mean abundance μ and the height F_a of the minimum abundance peak in the field data. The height is the number of species in the minimum abundance category a, whether an integer or a fractional number, as explained in the previous chapter. Transfer equations (Appendix A.2.8) taking μ and F_a as input may then be solved to yield ε and Δ as output. One may then plot the distribution developed by either method and compare it directly to the field histogram.

4.4.1   THE CHI-SQUARE TEST

When abundance data are compiled into a standard histogram, it becomes possible to compare the resulting shape with various theoretical proposals, including the J distribution. However, comparisons based on visual similarities can be misleading and somewhat dangerous. Not just some, but all of the abundance models developed prior to 1999 were evaluated by this method, which may be why there are so many.

An objective method of comparison is entirely quantitative and does not depend on subjective judgments. Karl Pearson (1900), who developed the chi-square goodness-of-fit test, used a special statistic to measure the difference between an actual abundance a_i and a corresponding theoretical abundance t_i:

$$d(a_i, t_i) = \frac{(a_i - t_i)^2}{t_i}.$$

The sum of such differences is called the *chi-square test statistic* and is compared with numbers in a chi-square table to determine to what degree the empirical data match the theoretical prediction. Each component in such a sum is considered as one "degree of freedom," meaning that it contributes freely and independently to the overall sum. However, each parameter of the theoretical distribution under test amounts to a restriction on how freely the terms may vary. Pearson therefore subtracted the number of such parameters from the number of terms in the sum, yielding the degrees of freedom for the test being carried out. The J distribution has two parameters, ε and Δ, so that a test involving 12 abundance categories, for example, would have 10 degrees of freedom.

When computing a chi-square test, there is a requirement that, when theoretical frequencies fall below 5.0 in a given category, such frequencies should be grouped with subsequent abundance categories until the sum comes to five or more. For example, if the theoretical abundances at 31 and 32 are 4.22 and 3.90, respectively, while the empirical abundances sum to 7, the corresponding term in the chi-square statistic would be

$$\frac{(7 - 8.12)^2}{8.12} = (0.15)$$

Further out in the abundance axis, the last empirical abundance category with a nonzero entry may well exceed Δ. Whether or not this happens, the last (grouped) category will use the sum of all remaining theoretical values to be compared with the sum of all remaining empirical values.

Under the null hypothesis, that the data observed actually arise from or follow the theoretical distribution in question, Pearson showed that chi-square test scores must themselves have a certain distribution, which he called the chi-square distribution, as shown in Figure 4.6. In other words, if one carried out 100 chi-square goodness-of-fit

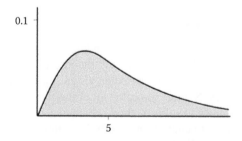

FIGURE 4.6   The chi-square distribution with five degrees of freedom.

tests (with 5 degrees of freedom) on data that one knew beforehand originated with the same J distribution (or any other distribution, for that matter), one would find that the test scores, when compiled into a histogram, would tend to match this outline. As one can see from the tail of the distribution in Figure 4.6, there will always be "outliers," instances of the distribution that fit rather poorly and so give rise to a high score, possibly off the page in Figure 4.6. Such samples may even match another theoretical distribution much better, yet they may well have originated in the distribution at hand!

As normally applied, the chi-square test involves a null hypothesis that the empirical data arise from the theoretical curve being matched to it. The resulting score is compared to a table of critical values, numbers that mark the boundaries between acceptance and rejection of the null hypothesis at various levels of significance.

Suppose that a chi-square test is carried out at 8 degrees of freedom on two different sets of empirical data, with resulting scores of 6.732 and 2.451 when compared with a particular theoretical distribution. Table 4.2 contains the critical values for 8 degrees of freedom (Pearson and Hartley 1972).

The upper row contains a set of q values or probabilities of a correct rejection, while the lower row contains critical values, one for each level of rejection. If a score exceeds a particular critical value, then the fit is rejected, with the corresponding probability q of being right. Thus, a score 6.732, which is greater than the critical value 5.071, involves a rejection of the null hypothesis with a probability $q = 0.750$ of being right. This leaves open the possibility (probability 0.25) that the null hypothesis is actually correct. On the other hand, the null hypothesis would be "accepted" at the critical value of 7.344 for the score of 6.732, but here, the probability of a correct rejection drops to 0.500. In this instance, "acceptance" simply means nonrejection. It does not mean that the data under test actually originated with the theoretical distribution at hand, although it may have. In short, the chi-square test is asymmetrical in regard to acceptance versus rejection.

In normal use, the test score is compared with the critical value at the $q = 0.950$ (95%) level of rejection; one wants a fairly high probability of being right in rejecting a hypothesis. At this level, the score of 6.732 is greater than the critical value of 2.733 that corresponds to the 95% level and the null hypothesis is rejected, with a probability of 0.95 of being correct. However, the other test score of 2.451, being less than 2.733, is "accepted."

To find an optimum fit, one might start with the values for ε and Δ found by the transfer equations. Since the empirical and theoretical distributions will match at the lowest abundance category, the first step is to compare the empirical and theoretical values, Δ' and Δ, respectively. If $\Delta < \Delta'$, the Δ should be increased to match, the coefficient c recalculated, and the chi-square test run again. The fit may be better,

TABLE 4.2
Critical Values for the Chi-Square Distribution

q	0.995	0.990	0.975	0.950	0.900	0.750	0.500	0.250	0.100	0.050
8 df	1.344	1.647	2.180	2.733	3.489	5.071	7.344	10.22	13.36	15.51

but now the low abundance columns will no longer match, so ε must be changed in a direction that produces a match for that category. Next, the chi-square test must be run again to see how much improvement there has been in the test statistic. Of course, once ε is changed in the previous step, a new value for Δ will emerge and the previous steps may be repeated. Sooner, rather than later, one is apt to find that no further improvement is possible and the resulting parameter values represent the J distribution closest to the data. (See the program BestFit in Appendix A.3.3.)

4.4.2   THE KOLMOGOROV-SMIRNOV TEST

The Kolmogorov-Smirnov (K-S) test (Hays and Winkler 1971) is somewhat simpler than the chi-square test, as it does not employ sums of differences. It helps one to determine if two sets of data follow the same distribution. Normally, the datasets are both derived from the field, but the K-S test may be adapted to curve fitting, if one of the datasets is in fact derived from a theoretical distribution with particular parameter values.

If the two distributions have n categories, one first calculates the corresponding cumulative distributions by adding up, for each value of k, all the entries in the respective categories up to the kth. We will denote the respective cumulative values by $F(k)$ (the empirical data) and $G(k)$ (the corresponding theoretical values), respectively. The K-S statistic has a simple formula:

$$D = \max\left\{\left|F(k) - G(k)\right|; \ k = 1, 2, \ldots, n\right\}.$$

In other words, D is simply the maximum absolute difference (between cumulative values) that occurs over the entire range of abundance categories.

Used in hypothesis testing mode, the results of a K-S test may be compared to a table of critical values, as was the case with the chi-square test. If the test score exceeds the critical value for a confidence level of 95%, the hypothesis that the two sets of values arise from the same distribution is rejected. When rejected at this level, one interprets the outcome as follows: "The two sets of data fail to follow the same distribution, with a 5% probability of this assessment being wrong."

In a K-S table, there is a critical value for each possible size of sample; the larger the sample, the larger the critical value must be. Acceptance of the hypothesis has the same interpretation in the K-S test as it does in the chi-square test; acceptance does not actually imply that the hypothesis is true, as there may be many theoretical candidates that would produce better fits. To better ascertain the presence of a particular underlying distribution, one needs many (say 50 or more) sets of data, testing each in turn and compiling the results.

We will return to the K-S test in Chapter 9, where it is applied to what might be called "fossil J-curves," namely, taxonomic abundance distributions, where one replaces counts of individuals in the present theory by counts of a lower taxon, as distributed across a higher one. For example, in a given geographic region (possibly the entire planet) and within a given family, there might be 21 genera with only one species, 14 with two species, 5 genera with three species, and so on. In this extension of the theory, the lower taxon plays the role of individuals, while the higher taxon plays the role of species.

4.5 APPLICATION EXAMPLE: SAMPLE OVERLAP AND SIMILARITY

The determination of sample overlap has its uses in field studies of richness, as in the studies cited in the following. It is statistically impossible to provide an unbiased estimate of overlap without knowing the underlying distribution of species over abundances.

Given two samples of the same community, how many species appear in both samples? For example, in his classic study of the Savannah River, Cairns (1969) used sample overlap to determine the similarity of communities. While the degree of overlap between two unbiased samples must, in a statistical sense, reflect the degree of similarity between the respective communities, the degree of similarity does not have a straightforward interpretation. It turns out that two *identical* communities will typically produce overlaps in the 70%–80% range, with typical sample sizes. Thus, an overlap of 75%, far from indicating a 3/4 overlap, may indicate a near-identity between the respective communities.

The similarity index, as derived in the following analysis, was used by the author (Dewdney 2010) in a study of benthic microbiota in a slow-moving river. The index is based on the empirical shape of the species/abundance distribution in the samples themselves and could therefore be called "parameter-free." The combined samples described in the next paragraph could be replaced by a best-fit J distribution with some hope of performing even better, but the difference between the two approaches has yet to be tested. In any event, the distribution obtained from the combined samples had the hyperbolic shape, and it is not clear that the results would be much different.

Suppose one takes two samples S_1 and S_2 of sizes N_1 and N_2 (number of individual organisms) from a community and suppose that the ith species appears a_i times in the combined samples. The ratio $a_i/(N_1 + N_2)$ then yields an unbiased estimate, p_i, of the relative abundance of the ith species in the community. This represents an estimate of the probability that an individual of the ith species will appear if one draws a single organism from the area sampled.

It follows that the probability of this species *not* appearing in such a drawing must be $q_i = (1 - p_i)$. Therefore, the probability of this species not appearing in a sample of size N_1 is $q_i^{N_1}$ and the complementary probability,

$$1 - q_i^{N_1},$$

represents the probability that the species will appear at least once in a sample of size N_1. Consequently, the probability of the ith species appearing at least once in another sample, this one of size N_2, is

$$1 - q_i^{N_2},$$

and the probability of the ith species showing up in both samples is simply the product of the two expressions:

$$\left(1-q_i^{N_1}\right)\left(1-q_i^{N_2}\right).$$

The expected overlap of the community with itself would then be given by the formula

$$E(S_1,S_2) = \sum\left(1-q_i^{N_1}\right)\left(1-q_i^{N_2}\right), \tag{4.1}$$

where the summation is taken over the union of species in the two samples. Naturally, since the samples are taken from the same community, one would expect the numerical value of Equation 4.1 to be close to the actual overlap.

The same formula may now be applied to the case where the samples S_1 and S_2 are drawn from different communities, C_1 and C_2, respectively. In this case, the value of E will reflect what the overlap would be *if* the two communities were the same. In fact, to the extent that the communities are different, the observed overlap will fall below the expected value $E(S_1, S_2)$. Thus, the ratio of observed overlap to expected overlap gives a reasonable measure of the degree of real overlap of the communities themselves. Although one is tempted to call the resulting measure the "community overlap," something like that is meant by the term "community similarity" or, more simply, the "similarity index."

We define the *similarity index* for two samples, S_1 and S_2, by the formula

$$SI(S_1,S_2) = \frac{O(S_1,S_2)}{E(S_1,S_2)},$$

where $O(S_1, S_2)$ represents the observed overlap between S_1 and S_2, namely, the number of species they have in common. This index represents essentially the true overlap normalized by the expected overlap to give a meaningful, full-scale (0 to 100, when expressed as a percentage) *estimate* of the degree of overlap of the respective communities.

In the microbial study just cited, the following overlaps between samples drawn from different sites were observed. The sites were labeled with a T (transect) code number, as shown in Tables 4.3 to 4.5. Table 4.3 shows the raw overlap figures, the number of species in common between the pairs of samples indicated by the table entry. Thus, the sample from Transect T5 and the sample from Transect T7A had 25 species in common.

Note that in Table 4.4, the expected numbers of common species are generally higher than the raw overlaps shown in Table 4.3, owing to the fact that the communities are different, whereas the expected overlaps are calculated on the assumption that the communities are the same.

TABLE 4.3
Raw Overlap Counts between Samples

Counts	T5	T6A	T6B	T7A	T7B
T4	26	22	28	37	33
T5		17	19	25	21
T6A			18	23	19
T6B				27	28
T7A					50

TABLE 4.4
Expected Overlap between Samples

Expected	T5	T6A	T6B	T7A	T7B
T4	29.9	35.9	38.8	56.0	58.9
T5		24.3	26.3	38.9	41.0
T6A			29.1	44.4	47.6
T6B				47.6	50.3
T7A					57.6

TABLE 4.5
Similarity Indices for All Pairs of Samples

%	T5	T6A	T6B	T7A	T7B
T4	87.0	61.3	72.2	67.9	56.0
T5		70.0	72.2	64.3	51.2
T6A			61.9	51.8	39.9
T6B				56.7	55.7
T7A					65.8

In Table 4.5, the similarity indices range from 39.9 to 87.0. It is permissible to interpret these as percentages, as in claiming that T4 and T5 are 87.0% similar. It is consistent with the high degree of similarity between T4 and T5 that the similarity between either the sample and the remaining ones should be relatively close together.

The similarity index technique was applied not only to different communities at the same time but also to the same community at different times, yielding a measure of change in the community over the period in question.

5 Predictions from Data

In the context of this chapter, "prediction" means the statistically accurate sample-based estimation of the richness of communities. It also means the cross-validation of parameters with each other as a check on the accuracy of the J distribution as a descriptor of samples.

In the first section, I show how the J distribution predicts the maximum abundance accurately (in the statistical sense) on the basis of other parameters, such as the mean abundance and the height of the initial peak (at lowest abundance). I also compare values of the maximum abundance Δ with the predictions of that parameter. In subsequent sections, I take a closer look at the sampling process, explaining why no richness estimation process can succeed without knowing both the sample intensity and the underlying distribution. I then illustrate this requirement with a brief review of some richness estimation methods already in use.

The second half of the chapter describes and compares two procedures for estimating community richness on the basis of sample richness in the context of the J distribution. Computer experiments make it possible to determine not only the relative effectiveness of the procedures but also the contributions to sample variance from irregularities present in the community being sampled, as well as those contributed by the sampling process itself. Distinguishing the two sources of variation, as well as their influence on the sample, is a crucial step on the way to a fuller understanding of the sampling process in its entirety.

Ideally, predictions from theory, at least numerical ones, should come equipped with error bounds so that biologists who make such predictions have a reliable estimate of the uncertainty that accompanies the predictions. Ecologists have been slow in coming to the realization enunciated by Doak et al. (2008): "Over the last decade, there has been increasing recognition that ecological predictions must be advanced with clear statements of their uncertainty." Such error bounds are a basic requirement of what I have called exact ecology. In the first section of this chapter, I show how to derive error estimates based on interval statistics arising from samples.

5.1 PREDICTING MAXIMUM ABUNDANCE

The two parameters of the J distribution, ε and δ, may be estimated on the basis of the mean μ of a sample and its initial peak F_a. The latter quantities have a functional relationship with the first two, being mathematically determined by them via the transfer equations (Appendix A.2.8). These equations were used extensively in the meta-study as reported in Chapter 8. With the sample mean μ and F_a as input, the transfer equations produce values of ε and δ that correspond to F_a (or, more commonly, F_1) and μ. The parameter ε is closely related to F_1 via R but not readily discernible in a sample histogram. The parameter Δ, on the other hand, is directly visible, in a sense, as the largest abundance Δ' in the sample; the computed value of Δ will be a good estimator of Δ' if the hyperbolic theory is correct, so it forms a relevant test for the theory.

How well does the derived value of Δ estimate Δ'? Very well indeed if the ratio Δ'/Δ, expressed as a percentage, equals 100.0%. More precisely, over a great many samples, how close does Δ get to Δ', on average? The average ratio Δ'/Δ in the collection of 125 meta-study biosurveys turns out to be 101.9, well within the expected range of possibilities for a correct theory and very close to the most likely value of 100.0 in relation to the standard deviation (34.6) of the ratios. The corresponding 95% confidence interval is [95.8, 108.0], with 100.0 very close to the middle.

While the foregoing result is entirely consistent with the prediction of the hyperbolic theory of Δ as the average maximum abundance, it must be remarked that actual abundances Δ' occurred more frequently (78 times) below 100.0 than above it (47 times). This is due to the fact that, while the size of ratios below 100.0 is limited, the size above it is not. In other words, while fewer ratios exceeded 100.0, they tended to be larger. The distribution of the percentages so averaged would probably resemble a left-skewed normal distribution. The result was expected and consistent with the interpretation of the J distribution.

These considerations concern only samples, of course. The maximum abundance in the *community being sampled* is another matter. The simplest form of prediction for the community Δ on the basis of the sample maximum abundance Δ' is to use the sample intensity r, namely as $\Delta = \Delta'/r$. Unfortunately, the high variance in the 125 Δ-ratios of the meta-study (Chapter 8) points to a rather wide margin of error inherent in such estimates. For example, in spite of a tendency for the estimate of Δ to be accurate in a statistical sense, an analysis of the 125 Δ-ratios indicates that such an estimate will lie within 10% of the true value only 17% of the time. If one allows a 30% error, the Δ-estimate will lie within 30% of the true value about 55% of the time. However, the project of obtaining more accurate versions of this estimate does not end with this observation. For example, repeated samples, although labor intensive, would result in an average value of the Δ-estimate to which interval statistics could then be applied with improved precision.

5.2 PREDICTING SPECIES RICHNESS

In the remainder of this chapter, the reader will encounter two epsilons and two deltas—those of the community and those of the sample, respectively. The community parameters will be indicated by ε and Δ, while those of the sample will be indicated by ε' and Δ', as in the practice followed above.

Suppose that a community C of N organisms inhabits an area or region A and that a biologist wishes to know the number R of species in C, based on its samples. As will be seen shortly, he or she must know both the intensity r of the sample, as well as the distribution underlying the community in order to have any hope of a realistic estimate for the number R.

I will present counterexamples to the possibility that such knowledge is not necessary to the project of determining R. The relative ease with which these counterexamples are produced is a strong indication that the project is not even approximately practical.

5.2.1   COUNTEREXAMPLE 1

The first counterexample shows that knowing the distribution underlying the community is critical to the determination of R, as shown in Figure 5.1. It involves two rather different artificial communities, C_1 and C_2, that have the same number N of individuals but with greatly differing numbers of species. Clearly, for C_1, R = 10, but for C_2, R = 20. Yet, when sampled, C_1 and C_2 yield the same number of species, at least most of the time.

Let C_1 have the univoltine distribution with 11 species, all having abundance 10, and let C_2 be the uniform community with 20 species, two in each abundance category, from 1 to 10. One distribution is a vertical spike; the other is long and perfectly flat. It is easy to verify that N = 110 in both cases so that, when sampled at intensity r = 0.1, the sample size would tend to be 11 in both cases.

For each species in C_1, the probability of *not* being chosen on a given occasion is clearly

$$\left(1 - \frac{1}{11}\right) = 0.901.$$

The probability of not being chosen 11 times in a row is the compound probability,

$$0.901^{11} = 0.318.$$

Since there are 11 species, the total contribution to their nonappearance, so to speak, must be

$$11 \times 0.318 = 3.494,$$

yielding a total expected number of species of 11 − 3.494 = 7.506.

One may apply the same analysis, albeit with a little more work, to the community C_2 as follows. The probability of a species of abundance k *not* showing up in the sample is $\left(1 - \frac{k}{110}\right)$. After 11 observations, the probability has dropped to

$$\left(1 - \frac{k}{110}\right)^{11}.$$

If one now merely adds up the contribution to nonappearances for each of the other abundances, the calculation takes only a few minutes:

$$\sum 2\left(1 - \frac{k}{110}\right)^{11} = 12.363.$$

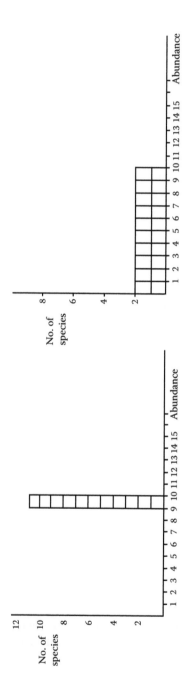

FIGURE 5.1 Two rather different communities.

In this case, the expected number of species in the sample must be 20 − 12.363 = 7.637. In other words, the two communities would yield the same number of species, on average, when sampled. Without knowledge of the distribution prevailing in C, the field biologist may make any prediction he or she likes but will, under one scenario or the other, be wrong.

It must be remarked that the shape of the samples of the two communities would differ somewhat, with C_1 having a higher initial peak than C_2 and declining somewhat more slowly, both petering out by the fifth or sixth abundance category.

With a sampling intensity and appropriate distribution in hand, the next step in estimating the richness R of the community is to reconstruct C by estimating its parameters. Before presenting the two main techniques of richness estimation based on the hyperbolic theory, I will review some estimation schemes that have already appeared in the literature.

5.2.2 INADEQUACIES IN CURRENT METHODS OF ESTIMATION

The natural interest in "biodiversity" in general and the need to know, specifically the richness of communities, have led ecological theorists to seek methods of estimating richness in communities. The methods have proliferated, just as have the proposals for species distributions, as explained in Section 1.2, as well as the definitions of biodiversity in Section 3.3. The proliferation of methods takes place, as in the other areas, against a background that involves two central problems that presently face the methodology used in the past: (a) a lack of adequate contact of theory with data and (b) a failure to recognize that "nonparametric" approaches, those that fail to consider the distribution underlying the community as a whole, cannot succeed in any meaningful way, as illustrated in this section.

The following review is hardly that. I have selected several of the more widely used methods to make the point just mentioned. Other studies could have been substituted for these without any change in the conclusions. The ultimate implication of this section is that none of the methods examined in the following sections could be called "exact" in the spirit of this book. Every method of which I am aware can only be called "approximate" and, indeed, prone to making false predictions.

Previous attempts to construct formulas or procedures for determining the number of species began with R. A. Fisher's derivation of the log-series distribution from the negative binomial (Fisher et al. 1943). Subsequent developments (see Chao et al. 2015 for a review of methods) were heavily influenced by this seminal paper, as we shall see.

5.2.2.1 The Fisher-Corbet-Williams Method

In deriving his species estimation formulas, Fisher assumed that the parent community follows the normal ("Eulerian") distribution. This assumption stands in direct contradiction to the theorem that the species-abundance distribution of a sample must follow the distribution of the community it was drawn from (Section 3.5). Field histograms never resemble the normal distribution and, thanks to the general theory of sampling, it can be said with some confidence that community distributions are never normal, nor even approximately so.

In any event, Fisher developed two equations that allow one to fit a log-series curve to a field histogram:

$$R' = -\alpha \ln(1 - x)$$

and

$$N' = \frac{\alpha x}{1 - x}.$$

Here, as elsewhere in this book, R' and N' represent the number of species and individuals, respectively, in the sample. The parameters α and x are explained in Appendix A.2.7. Here, we observe simply that when the sample values for R' and N' are substituted into these equations, corresponding values for the parameters α and x can be extracted by solving the equations. This was a troublesome matter in the 1930s, when this work was done, but computers today may solve these equations instantly. With values for α and x in hand, one has the theoretical form of the distribution thought by the authors to describe the sample.

Fisher went on to claim that, since the parameters α and x seem to remain constant over different sizes of sample from the same community, the relation between R' and N' can be extrapolated via these parameters, as embodied in Equation 5.1. This conclusion is largely conjectural, owing to the fact that only a few samples of animal communities were examined in any detail. In any case, one may combine the two equations above to obtain,

$$R' = -\alpha \ln\left(\frac{\alpha x}{N'}\right). \tag{5.1}$$

Here, as N' increases, the quantity $\alpha x/N'$ decreases. However, because the argument of the logarithm lies between 0 and 1, the logarithm itself is negative and increasingly so as N' increases. The minus sign thus makes the entire expression positive.

Essentially, the expected number of species increases as the logarithm of the sample size. It is not clear how far the authors thought the method might be pushed in moving toward an estimate for R, the richness of the community as a whole. Since Fisher thought the community had a normal distribution, he would probably hesitate to suggest such a role for his formula. In his view of the matter, the log-series shape of sample histograms would necessarily morph into normality as N' approached N (the size of the community), invalidating the method. But given the closeness of the log-series distribution to the J distribution, at least at the low-abundance end of the axes, Equation 5.1 might be used as a rough guide to the manner in which species accumulate as sample size grows. In the context of the Fisher, Corbet, and Williams study, it would have been more consistent to assign the log-series distribution to a role in communities, as well as samples. However, any log-series distribution for the community would have much lower values for α than prevailed in samples, so an extension of the method would have to incorporate decreasing values for α.

It might be added that the close resemblance of the log-series shape to that of the J distribution made it the prime candidate for comparison with the J distribution in the meta-study of Chapter 8.

5.2.2.2　The Goodman Statistic

Leo Goodman (1949) proposed the following estimator for the number of species in a community having N individuals and a distribution F′ of species over abundances in a sample. Goodman used an intermediate quantity S′ in his formulation, here simplified to

$$S' = N - \left(\frac{N}{n}\right) F'(2),$$

where n is the sample size and F′(2) is the number of species having abundance 2. Goodman's estimator R may then be defined as follows:

$$R = S', \text{ provided that } S' \geq \sum F'(i)$$
$$= \sum F'(i), \text{ provided that } S' < \sum F'(i),$$

where the summation is taken over all abundances appearing in the sample. It will be noted immediately that the summation can be replaced by R′, the number of species to show up in the sample, so that one has the much simpler recipe,

$$R = S', \text{ provided that } S' \geq R'$$
$$= R', \text{ provided that } S' < R'.$$

Goodman claimed that although the previous statistic is unbiased, it may be applied only when the sample size n equals or exceeds the largest population in the community. In Example 1, described earlier, two markedly different communities C_1 and C_2 shared the same value of N. Moreover, when a sample of size 11 was taken of both communities, the same number of species tended to appear in the sample, namely 7.

Using the program SampSim (see Appendix A.3.2), I sampled both communities 100 times at intensity r = 0.1 and obtained the following average values for F′(2) and S′ (Table 5.1).

Since S′ > R′ in both cases, the Goodman formula estimates 88.9 species in C_1 and 94.1 species in C_2, both estimates being rather far off the actual richness values of 11 and 20, respectively. In fairness to Goodman, it must be mentioned that his

TABLE 5.1

Values of F'(2) for the Two Samples

Community	F'(2)	S'
C_1	2.11	88.9
C_2	1.59	94.1

formula applies only to samples without replacement. On the other hand, one may suspect that the communities C_1 and C_2 may be scaled up to the point where it makes little difference whether one samples with or without replacement. Any failures in this method would clearly be due to not taking into account the distribution that prevails in the community.

5.2.2.3   The Jackknife Estimator

When a survey is taken of a population P of an animal species, a field biologist may capture an individual, mark it in some manner, then release it. The aim of this particular survey is to count the population. Allowing for the possibility that some individuals will be less prone to capture than others, one admits varying probabilities of capture, say $p_1, p_2,..., p_m$, where m is the population size.

A well-known method due to Burnham and Overton (1979) has been employed under these conditions to estimate the population P. Assuming that the individual capture probabilities are unknown, the method assigns a probability drawn at random from the interval [0, 1], the distribution being uniform.

The authors noticed that the problem as stated could be reinterpreted in the context of sampling a community for its species. The similarity can be nicely illustrated by the standard urn model of probability. An urn contains N balls marked with various numbers that are distributed among the balls according to the uniform random distribution just mentioned: if n(i) represents the number of balls receiving the numeric label i, then let F(k) represent the number of labels i that have k balls in their respective groups.

The labeling can be arranged so that for each i,

$$\frac{n(i)}{N} = p_i.$$

Clearly, if one samples the urn with replacement at each turn, the probability of drawing a ball from the ith numeric class equals the probability of capturing the ith individual in the population survey just described.

However, it may also be interpreted as the problem of sampling a community of species, with the ith species being represented by all the balls labeled i. If one were to make a histogram of this particular community, it would look basically level, with the usual statistical variations. The jackknife method is applied not only to the mark/recapture problem but also to the problem of sampling communities of species. However, in the latter context, it clearly assumes a uniform distribution within the community, and this cannot be correct, since it contradicts the theory of sampling (see Section 3.5).

5.2.2.4 The Bootstrap Method

Perhaps the main motivation behind nonparametric methods is to avoid having to know the distribution that prevails in the community, the underlying notion being that this information is present to some degree in the sample. The idea is clever, but subject to the degree to which the information is "present." As will be shown here, the bootstrap method has problems with certain distributions, a circumstance that undermines the idea of it being nonparametric.

The method, due to Smith and van Belle (1984), uses a sample of n individuals to produce new samples by sampling the sample, so to speak. Since the sampling takes place (with replacement) at intensity 1.0, it amounts to a perturbation of the sample, as defined in Section 8.1.1. If the ith species shows up p_i times (relative frequency) in the perturbed sample, the following *bootstrap formula* is claimed to estimate community richness:

$$R = R' + \sum (1 - p_i)^n.$$

Here, the summation is taken over the species present in the sample and R' is the richness of the sample. The process can be repeated, with a final estimate calculated as the average of estimates thus developed.

The term $(1 - p_i)$ represents the probability that the ith species does not show up in a single drawing, whereas the term $(1 - p_i)^n$ represents the probability that it does not show up in the sample at all. It will be noted that the same kind of probabilistic calculation was employed in the example that heads this section.

If p_i is small enough, the term $(1 - p_i)^n$ may not be negligible and the contributions from all such terms to the summation would be substantial. In other words, according to the previous formula, the summation itself amounts to an estimate of the number of species that did not appear in the sample; the low-abundance species are used to estimate the number of missing species. For a sample size of 100, for example, a species of abundance 1, 2, or 3 in the sample will contribute the quantities 0.366, 0.133, or 0.048, respectively, the contributions from higher abundances decaying rapidly to zero.

By not taking intensity into account, the method fails, as the following counterexample shows.

5.2.3 COUNTEREXAMPLE 2

Let C_1 be the hyperbolic community J[2.0, 1000] × 50 and let C_2 be another, smaller hyperbolic community that happens to be an idealized sample of C_1 taken at intensity r = 0.1. When applied to C_1, the program Samplesim (see Appendix A.3.2) will produce such a sample by averaging over many samples, all taken at the same intensity. This sample may also be considered as a hyperbolic community in its own right and turns out to have an average of 36.2 species. The same program can now be used to sample C_2 in the same manner, but this time at intensity $r = 0.5$. The final result is a sample that turns out to have an average of 29.5 species.

The point of this example is that the resulting (average) sample could have been obtained by sampling C_1 at intensity 0.05 or by sampling C_2 at intensity 0.5. The sample has 29.5 species, while C_1 has 50 and C_2 has 36.2 species. The bootstrap method cannot make both predictions simultaneously, nor is it likely to make either, since there is a potential infinity of examples, all based on the same original distribution. Moreover, the same kind of example can be produced for any starting distribution one likes. I have used the J distribution for this example, since my software is geared to it. By using average samples, I have avoided the possibility of producing an example on which the bootstrap method fails by coincidence. The averages used in the bootstrap method itself do not appear to improve the situation.

5.3 PARAMETRIC VERSUS NONPARAMETRIC APPROACHES

A distinction between "parametric" and "nonparametric" approaches to species abundance estimation is frequently made in the population biology literature. How would the distinction be made for the approaches outlined in the previous section?

The Fisher-Corbet-Williams method assumes that the community follows the normal distribution, while the sample is fitted with a log-series distribution. As already remarked, these two assumptions directly contradict the sampling theorem in Section 3.5. In any case, the Fisher-Corbet-Williams method would be described as parametric as a normal distribution is assumed, a distribution with parameters.

The Goodman statistic is nonparametric, as no particular distribution is posited for the community—or any sample of it.

The Jacknife estimator appears, at first glance, to be nonparametric, since it uses no parameters directly. However, as remarked earlier, it carries with it the implicit assumption of a distribution in which all species have the same abundance. The shape of such a distribution is univoltine, a single spike, or column in its histogram.

The Bootstrap Method is clearly nonparametric and relies heavily on the shape of the sample histogram to assess the source community. Although explicitly nonparametric, the method is implicitly parametric; if samples tend to all be manifestations of a single, overriding distribution, having a J-shape in practice, then the method depends implicitly on the subject of this monograph. The same thing is true of the last method reviewed earlier.

A recent nonparametric approach to abundance prediction for communities (Chao et al. 2015) uses rank abundance (see Section 2.2.3) as the setting for the method described by the authors. "Most previous approaches (e.g., Dewdney 2000, Green and Plotkin 2007) are based on a parametric assumption about the SAD [species-abundance distribution] of the entire assemblage." The authors of this most recent article apparently failed to notice that the J distribution was hardly an "assumption." In any case, the method proposed uses the sample abundances of a community to form an estimate of the abundances in the community. The method is based on the Good-Turing method of species estimation (Good 1953), which assumes a normal distribution in the fluctuations encountered by the sampling process. In any case, this recent proposal encounters problems since it does not use sampling intensity as part of the estimation method. As shown in the previous section, it will inevitably make richness estimates that are wildly off. As already pointed out earlier in this chapter, the same sample can arise from communities that differ by an order of magnitude in richness; there is simply no way to distinguish a low-intensity sample of a large community from a high-intensity sample of a small one.

5.4   EXACT ESTIMATION METHODS

In spite of the critical nature of the distribution underlying abundances in a community, the following sampling formula plays no favorites but applies to all possible (continuous) distributions g that might lurk in the community. As explained in Section 3.5, the Pielou transformation represents an accurate template for the sampling process. It expresses the number F of species of abundance k to show up in a sample of intensity r:

$$F(k) = \int_{0}^{\infty} \left(\frac{e^{-rx}(rx)^k}{k!} \right) g(x)\, dx, \tag{5.2}$$

where $g(x)$ is the distribution of abundances in the community being sampled. In the context of this book, the function g will always be the J distribution. The Pielou transform is based on standard sampling theory as explained by Feller (1968). The formula is an extremely close approximation (error less than 0.1%) to the hypergeometric formula on which the theorem is based. Indeed, the formula is also implicitly present (although not explicitly used) in one of the earliest papers on sampling in the ecological literature (Fisher et al. 1943).

It is not difficult to embed Equation 5.2 in a computer program that takes an arbitrary (community) distribution as input and produces expected samples for every possible value of the sample intensity r. (See a description of the program CommRich in Appendix A.3.2.) These outputs turn out to be essentially identical with the outputs of another program that simulates the sampling process itself in a statistically precise way. The program produces the expected number of species for each abundance category in a sample of a size implied by r. This program is used to evaluate two methods for estimating the richness of a community from its samples. In doing so, it illustrates a general method of proceeding in estimation research.

Both methods employ the same basic procedure of cycling back and forth between sample data and a test community $J[\varepsilon, \Delta] \times R$, as shown in Figure 5.2. The community is sampled iteratively, always at the same given intensity r, but each time with different values for ε, Δ, and R, until a certain combination of the parameter values produces a sample that most closely matches the sample in hand. The iterative process works best when the user employs the method of steepest descent, always moving in the direction of an improved match and arriving at a best match after a finite number of steps.

The sample drawn from the community at each stage of the process is an *expected sample* arrived at via the Pielou transform, as in Figure 5.2. In the expected sample,

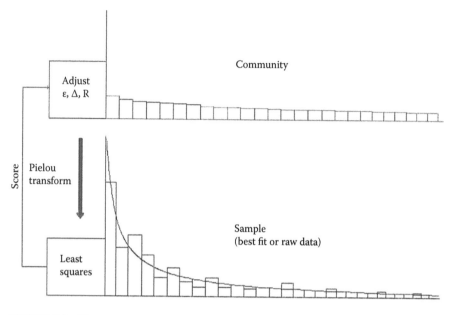

FIGURE 5.2 The estimation process converges to a best estimate for R.

each abundance category is inhabited by a fractional number of species, such as 12.64 or 0.42, etc. These numbers have the J distribution and they may be compared directly with their counterparts in the sample at hand. In the two-step method, the sample is computed as the best fit to the field histogram. In the one-step method, the sample is the histogram itself. The least squares function provides the vehicle for the comparison, where F″(k) is the expected value for the number of species in the kth abundance category and F′(k) is the actual number of species in the sample at hand (whether raw or computed) that has that abundance:

$$\text{difference} = \sum \left(F''(k) - F'(k) \right)^2.$$

The basic cycle may be described as follows:

1. Assign initial values to the parameters ε and Δ for the community, as well as the community richness, R.
2. Use the Pielou transform to produce the expected sample of this community.
3. Compare the resulting theoretical sample with the one in hand via the least squares measure.
4. If the match is worse than before, reset the most recently changed value of ε, Δ, or R in the opposite direction. If the match is better, continue as before.

Presently, the entire process is embedded in one of two computer programs, depending on the method, and the complete richness estimation process may take anywhere from 2 to 200 cycles to complete. In other words, a human must execute the search algorithm, a process that can itself be automated, cutting the estimation time from an hour or two down to less than a second. The algorithm itself would systematically cycle through ε, Δ, and R, changing each until no further improvement is seen, then switching to the next parameter. At no point do the corresponding parameter values for the sample at hand play any role. The sample richness R′ plays an implicit role, however, through the values of the sample function F′ at each abundance category.

5.4.1 THE TWO-STEP METHOD WITH AN EXAMPLE

The first procedure described here is called the *two-step method*. It proceeds in two main steps:

Step 1. Find a best fit for the sample histogram with the J distribution. The program called BestFit does this, taking the sample histogram as input, then comparing these data with the numbers generated from a theoretical (hyperbolic) sample distribution with (sample) parameter values input by the user of the program. The values of ε' and Δ' thus arrived at can be varied systematically over the parameter space to discover a global minimum in solution space. In most cases, the method of steepest descent finds the minimum without having to search the entire space. The measure of fit is the chi-square score divided by the number of degrees of freedom, as determined by the program. This method of scoring helps to minimize jumps in score values that would otherwise result when the program changes the number of degrees of freedom.

Step 2. One then inputs the best fit parameter values of the J distribution into the program CommRich, along with the sample intensity estimate, r, made by the biologist. The user then conducts a directed search through solution space by systematically varying the community parameter values ε and Δ, as well as the community richness R, as described previously; for each set of values thus arrived at, the program computes values for the expected sample and compares the theoretical sample with the best fit curve from step 1 using the least squares formula as a measure of similarity. The underlying algorithm uses the smallest least squares score found so far as the basis for further improvements in the score. Any change in a parameter value that leads to an improved score is adopted as the starting point for the next cycle. The change is not selected arbitrarily, but on the basis of producing the greatest improvement of the score, as it steadily descends to lower values. At the end of the convergence process, one reads off not only statistically accurate estimates for ε and Δ in the community but also its richness, R, as a byproduct of the process.

Time to convergence during either fitting process depends strongly on the starting parameter values. But a form of binary search may be employed that speeds the process up, completing in a time that is proportional to the logarithm of the size of the parameter space being searched.

An example of the method in action is provided by data sent to me by M. G. M. Jansen, a Dutch biologist who had been conducting an extensive sampling program for Lepidoptera inhabiting coastal salt marshes in the Netherlands. Table 5.2 displays the data from one of Jansen's samples as whole numbers. Each cell of the table under the heading "no. spp." also contains the corresponding number of species predicted for the corresponding abundance as fractional numbers. The table shows observed abundances for some 45 species, the remainder having abundances 31, 32, 41, 67, 103, 103, 1121, and 2073.

Running the program BestFit on these data, I found a best fit for the chi-square measure at parameter values $\varepsilon' = 0.358$ and $\Delta' = 124.7$. The resulting chi-square score was 0.5372 at 6 degrees of freedom. The optimum fit was obtained by starting at initial values of $\varepsilon' = 0.5$ and $\Delta' = 140.0$ and continuing to adjust the values of the two parameters, steadily decreasing the chi-square score until no further improvement was possible, as described in Section 5.4.1. The procedure required 22 steps in this case.

With these values of ε' and Δ' as input, the program CommRich produced a theoretical version of the sample, storing it in an array for the ensuing cycle of comparisons. The first 30 parallel values arise from the distribution $J[0.358, 124.7] \times 54$.

Jansen's estimate of sample intensity, namely, $r = 0.0017$ (0.17% of the community sampled), was inputted to the program. I then followed the method of steepest descent, starting at initial guesses of $\varepsilon' = 1.0$, $\Delta' = 2200.0$, and $R = 100$. It took some 60 steps to arrive at values for all three parameters that jointly minimized the least squares measure. The values were $\varepsilon' = 2.50$, $\Delta' = 2215.0$, and $R = 127$.

Jansen thought the R-estimate rather high. He would have estimated the number to be closer to 100, based on intuition about salt marsh lepidopteran abundance in general. However, he had shown some uncertainty about his estimate for the intensity, r, of his sample. The reasons for his uncertainty lay in the nature of his subject. The salt marsh under study had some variety in its vegetative structure and some species seem to be more abundant than samples would indicate, as though sample intensities were not uniform. In any case, for a higher value of r,

TABLE 5.2
Sample Abundances vs. Predicted Ones

Abund.	No. spp.		Abund.	No. spp.		Abund.	No. spp.	
1	15	15.24	11	1	0.87	21	1	0.42
2	7	5.75	12	0	0.79	22	0	0.40
3	3	3.60	13	1	0.73	23	0	0.38
4	3	2.62	14	0	0.67	24	0	0.36
5	2	2.05	15	0	0.62	25	1	0.34
6	1	1.68	16	2	0.58	26	0	0.33
7	2	1.43	17	0	0.54	27	0	0.31
8	2	1.23	18	0	0.50	28	0	0.30
9	1	1.09	19	0	0.47	29	0	0.29
10	1	0.97	20	1	0.45	30	1	0.28

the estimate for R would have been lower. On the other hand, field biologists are often surprised at communities that turn out to be more speciose than they thought (Whitfield 2003).

How good is the R-estimate? The answer is provided by a series of experiments described in the next section. In the case at hand, the estimated number of species in Jansen's lepidopteran community was 127 species, give or take 6% (8 species) 95% of the time. In other words, with probability 0.95, the lepidopteran community had between 119 and 135 species—at least if the r value supplied by Jansen was approximately correct. If his estimate of the sampling intensity r was low, however, and if the actual value was 0.0020, for example, the R-estimate would drop to 106 species.

All of this raises the obvious question of how we can ever know the number of species in a community that is under study. The simple answer is that unless we sample everything, we cannot. However, the methods described in this book will produce estimates with a statistical accuracy that, in the author's opinion, cannot be improved. Two kinds of communities are used as test beds for the estimation procedures. One kind is purely theoretical, being a strict J distribution with the parameter values currently under test. The other kind is a perturbation of such a community, resulting in a distribution that I will claim to be typical of the distributions that prevail in real communities.

5.5   EXPERIMENTAL EVALUATION OF METHODS

The following experiments are not used to prove that richness estimation methods work. In the context of the hyperbolic theory, they are already known to work. The experiments here serve merely to (a) illustrate the accuracy of the method and (b) provide interval statistics (based on variance of the samples) so that reliable error estimates of accuracy can be made. This means that each estimate made by this method will have an accompanying error term.

A second set of experiments was piggybacked onto the first set. By using two versions of the community being sampled, it was possible to arrive at preliminary estimates of the relative importance of variation in community abundances versus variation in those of the sample. Two forms of the community being sampled were used. In one of these, the community was represented by a smooth theoretical function. In terms of hyperbolic theory, such a community will have zero variation. The second form of community was a perturbed version of the first. A total of 75 tests were carried out, each involving a random sample drawn from an H community using the computer program called SampleSim.

5.5.1   THE TWO-STEP METHOD

In the first experiment, 25 random samples were drawn from an idealized community J[2.0, 3000.0] × 50. In the second experiment, 50 random samples were drawn from randomly perturbed versions of that community. These were obtained by simply sampling (with replacement) the community at intensity 1.0. The results of these experiments are summarized in Tables 5.3 and 5.4. In each case, the average estimated richness appears under the heading "Mean."

TABLE 5.3

Richness Estimates and Error Terms for Idealized Community

r Value	No. Tests	Sample Size	Mean	SD	Error
0.005	10	68	50.47	7.02	8.3%
0.010	15	136	49.50	3.47	5.0%

TABLE 5.4

Richness Estimates and Error Terms for the Perturbed Community

r Value	No. Tests	Sample Size	Mean	SD	Error
0.005	12	68	46.9	10.4	13.6%
0.010	15	136	50.3	10.6	12.5%
0.020	14	271	49.4	5.7	6.8%
0.020	10	387	51.4	3.9	5.6%

The two average richness predictions that appear in Table 5.3 are closer to the ideal average of 50.0 than are the average predictions shown in Table 5.4. This is due to the data being less variable in the first table, given that samples came from a smooth, idealized community. At the same time, we observe that in both tables, the error term for intensity 0.005 is larger than the error term for intensity 0.010. This is due to the samples being twice as large in the second case.

The results displayed in Table 5.4 show a similar trend in accuracy; the error estimate declines in step with higher sampling ratios. The last set of (10) samples in Table 5.4 was drawn from a much "larger" community, a perturbed version of J[10.0, 4000.0] × 50. This not only illustrates the method's wide applicability but also provides an extra data point when plotting estimation error as a function of sample size (see Section 5.5). The average richness estimate for all four tests was 49.5, acceptably close to 50.

5.5.2   THE ONE-STEP METHOD

In this method, the second step of the two-step method is used in the same manner, but with a different input—the raw data itself, instead of the best fit to it. There is, therefore, no first step to speak of. Tests of the one-step method parallel those of the two-step method, with the first series of experiments performed on the idealized community J[2.0, 3000] × 50. The remaining tests focus on perturbed versions of the same community. The series two tests thus assess the accuracy of the one-step method, enabling a direct comparison of the results.

TABLE 5.5
Richness and Error Estimates for the Idealized
Community

r Value	No. Tests	Sample Size	Mean	SD	Error
0.005	10	68	53.0	8.9	10.4%
0.010	15	136	49.1	4.0	5.7%

TABLE 5.6
Richness and Error Estimates for the Perturbed
Communities

r Value	No. Tests	Sample Size	Mean	SD	Error
0.005	14	68	52.0	10.9	13.2%
0.010	14	136	51.9	10.3	12.6%
0.020	14	271	51.4	7.0	8.4%
0.020	10	387	51.4	3.4	5.7%

Results of the first series of experiments on the idealized community J[2.0, 3000] × 50 are summarized in Table 5.5, while those for perturbed versions appear in Table 5.6. Apart from what appear to be significantly higher error terms in the one-step method, there is little to choose between the two methods. Given that the mean richness estimate is independent of intensity, one can arrive at a slightly more refined richness estimate of 49.98% for the two-step method and 51.05% for the one-step method, both applying to the idealized community.

The foregoing results arise from an idealized community, of course. The second set of experiments involved not an idealized community, but a great many perturbations of it. In this case, the richness estimates averaged 49.5% for the two-step method and 51.7% for the one-step method. The second average seems a bit high, but the 50% target is well within the 8.4% error interval.

5.6   THE BEHAVIOR OF ERROR TERMS

Although more random communities could be sampled in order to obtain more precise data for plotting error as a function of sample size, the results of the previous section yield a reasonably close first approximation. Table 5.7 illustrates how expected errors (via the error curve in Figure 5.4) may be systematically derived from the experiments described in this chapter.

The error curve shown in Figure 5.3 is an optimum fit to the error data shown in Tables 5.4 to 5.7, using a function of the form $y = k/\sqrt{x}$, where x is the sample size. This is a standard application in statistics, error declining as the inverse square root of sample size.

TABLE 5.7

Practical Table for the Assessment of Errors in Richness Estimation

n	60	80	100	120	140	160	180	200	220	240	260	280
Error	15.9	13.9	12.4	11.4	10.5	9.9	9.4	8.9	8.5	8.1	7.9	7.6
n	300	320	340	360	380	400	420	440	460	480	500	520
Error	7.3	7.1	6.9	6.8	6.6	6.4	6.3	6.1	6.0	5.9	5.8	5.8

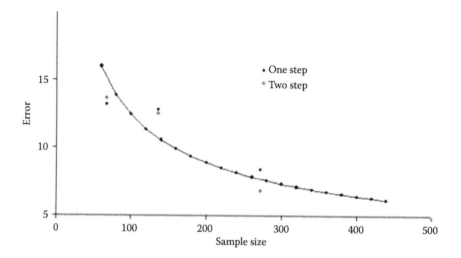

FIGURE 5.3   Richness estimation error as a function of sample size.

As will be seen in the next section, the percentage contribution to variance by the source community is relatively low for small sample sizes, with (pure) sample variance predominating. The resulting error curve is therefore the weighted sum of two functions, the weights slowly shifting from predominantly sample-influenced to community-influenced, as sample size increases. The approximation curve is reasonably close to the error data and, particularly for larger samples, may overestimate the error.

As can be seen from the relative positions of open (one-step) and closed (two-step) circles, there is little to choose between the two methods in respect of accuracy. The entries in Table 5.7 were read directly from the curve, in effect. The table, although useful in a limited way, serves only as an example until further experiments sharpen and extend these results.

5.7   ANALYSIS OF THE TWO SOURCES OF VARIANCE

As was pointed out earlier in this chapter, both the community and the sample contribute some variance to the final estimate of richness. The point of the Series 1 tests was to tease out the variance due to the sample alone. The results of Series 2

TABLE 5.8

Contributions to Variances in Methods From Smooth vs. Perturbed Communities

Ratio R	Sample Size	No. Tests	Two-Step: Variance	One-Step: Variance
		Idealized Communities		
0.005	68	10	49	79
0.010	136	15	12	16
		Perturbed Communities		
0.005	68	10	108	118
0.010	136	15	94	106

incorporate both sources of variation, and they are, of course, larger. Corresponding entries of Tables 5.7 and 5.8 may be compared to analyze the respective contributions at two intensities or, in this case, sample sizes.

Since the two sources of variation are completely independent, the variance in the sample may be calculated as a percentage of overall variance:

For the two-step method, the proportion (percentage) of independent sample variance is about 45% at sample size 68, dropping to roughly 13% at sample size 136. For the one-step method, the proportion falls from approximately 68% to 15%. It makes sense that the proportion should drop since larger samples tend to have less relative variation than do smaller ones for the same source community, whatever its state. On the other hand, the one-step method produces larger percentages across the board, the difference diminishing as the sample size increases. There is a clear tradeoff between the two methods. The one-step method is easier to carry out and enjoys much the same (long-term) accuracy as the two-step method, but it produces a more diffuse kind of prediction, for which the error term is significantly larger than that of the two-step method.

5.8   ASSESSING GENERAL POPULATION DECLINES

It may happen that, owing perhaps to some adverse environmental factors, there is a general population decline spread over a large group of organisms. For example, as this book was going to press, it was being reported by numerous entomologists around the world that a general decline in insect populations appeared to be in progress (Smith 2016). The problem addressed by this last section is how to assess the loss of species as a function of the loss of individuals. If a community of insects loses 20% of its individuals, how many species have been extirpated as a result?

To make such an assessment, it may be useful to look through the other end of the species telescope, so to speak. It will be assumed that in such general declines, species lose individuals at random and the total population thus extirpated has all the characteristics of a random sample, at least in a community context. Ironically, the remaining individuals also have the characteristics of a sample, albeit one with high intensity; the worse the decline, the smaller the remaining "sample" becomes.

So let a certain community with N individuals and R species suffer a decline to the point where N′ individuals and R′ species remain. The corresponding "sample intensity" turns out to be r = N′/N. If we knew R, the job of estimating R′ would be much easier.

In order to use the methods of this chapter in this situation, one must sample the community so affected (a sample within a "sample") and apply the one-step or two-step method to the sample in order to arrive at a reasonable estimate of R′, the richness of the remaining community, along with estimated values for parameters of the corresponding J distribution. At such a point, one may turn around and treat this information as a real sample, then apply the one- or two-step estimation methods to reconstruct the original, predecline community, obtaining in the process a reasonable estimate for R, the original richness. Written as a percentage, the ratio (R − R′)/R expresses the loss in terms that are readily understood. And of course, the J distribution has a role to play in answering the question. Such an approach might well suffer from even higher variances, but the scheme is definitely workable.

Another window on the species loss problem is opened by a different approach to analyzing its effect on the J distribution for a community. This time, we use the variable r not for the number of individuals gained in a sample but for the number lost to the community as the result of a decline. When a community loses a proportion r of individuals, it also loses species, but how many? A somewhat simplistic approach has suggested itself as this book was going to press; if a community loses a proportion r of its individual members and if the loss is distributed evenly over its species, then the abundance x of each species will decline to rx. By applying the same factor to the parameter ε, we may write an expression for the resulting distribution function as follows:

$$Rc(r/(x + r\varepsilon) - \delta/r).$$

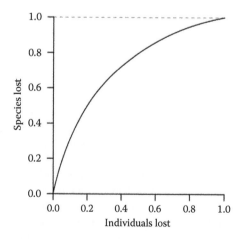

FIGURE 5.4 The proportion of species lost as a function of individuals lost.

Integrating this expression from rɛ to ɛ should yield a reasonable estimate of the number of species having abundances less than zero, so to speak, i.e., extirpated. The resulting integral contains the factor

$$r \ln((1+r)/r),$$

and Figure 5.4 shows a plot of the resulting shape. Interestingly, this particular approach to the problem results in a parameter-free estimate of lost species. If true, it would mean, for example, that a loss of 20% of individuals across the board would result in a loss of 51% of the species, a rather alarming result. It undoubtedly reflects the large number of species that already have low abundance.

If this finding is borne out, it would provide a fast way to estimate loss of species, once some solid data on the percentage losses of individuals can be obtained.

6 Extending the Sample

It is very easy to predict the effect of extended sampling on the parameter Δ; if the size of a sample doubles, so should the expected size of the largest population. However, it is not so easy to discover how the parameter ε tends to behave in this situation. The analysis in Section 6.1 indicates a linear relationship, with further work still required.

If continued sampling in a specific area results in samples with increasingly higher values of ε, it also results in ever fewer new species per sampling effort. Two additional applications of the hyperbolic theory tell us how many more species to expect if (a) the sampling intensity is increased by a given proportion over the same sampling area and (b) if the area being sampled is increased by a given proportion. The first application yields what is known as an accumulation curve, the complementary notion being called rarefaction, the tendency of ever-fewer species to show up. The second application yields a new species–area relationship. Both areas of inquiry have been intensively investigated, but all too often with somewhat arbitrary assumptions at the beginning. No new assumptions are added to the hyperbolic theory here.

6.1 THE EFFECT OF SAMPLING ON PARAMETERS

When one samples a community with intensity r, the natural question is how the sample parameters ε' and δ' of the sample are related to r. For the parameter δ', the situation is simple. A sample of intensity r will map the maximum abundance Δ in a community into a maximum abundance $\Delta' = r\Delta$ in the sample. In other words, in a well-taken sample, δ'/r is an unbiased estimator of δ'. The situation for ε may ultimately be as simple as this but is more difficult to elucidate. In order to develop an appropriate theoretical approach to the epsilon problem, the following formula applies the integral equation in Section 2.2.1. Owing to its intimate relationship with ε, we would propose to use the number $F(1)$ of species of unit abundance as the vehicle to tie together the quantities ε and r, the sampling intensity:

$$F(1) = Rc\left[\ln\left((1+\varepsilon)/(k-1)+\varepsilon\right) - \delta a\right],$$

where $k = 1$ and $a = 1$, so that

$$F(1) = Rc\left[\ln\left((1+\varepsilon)/\varepsilon\right) - \delta\right].$$

The parameter ε occurs three times in this formula, one of the occurrences being concealed within the constant c. The fully explicit formula for $F(1)$ turns out to be

$$F(1) = R(\ln(\Delta/e) - 1)\left[\ln\left(\varepsilon/(k-1)a+\varepsilon\right) - \delta a\right]. \tag{6.1}$$

The parameter r resides in the Pielou transform (Section 3.5, Equation 3.2). Specialized to k = 1, the transform would give us another expression for F(1). Setting the two expressions equal would (hopefully) produce an explicit, solvable equation that ties together ε and r. But it does not, so we must experiment in order to discover whether such a relationship might exist. In any case, the foregoing difficulty is left in place as an illustration of how mathematics drives the research direction in hyperbolic theory.

The following experiments, although minimal, seem to imply a linear relationship. The value of ε increases directly with the value of r. A total of 36 experiments were performed. Each involved four different distributions (representing communities) that were sampled at nine different intensities. The results were plotted, as in Figure 6.1. Eight of the intensities were concentrated in the range r = 0.005 to r = 0.012, with an additional value at r = 0.020 as a check. The communities had the form J[ε, Δ] × 50, with all four combinations of ε and Δ present, respectively: (a) 10.0 and 4000, (b) 10.0 and 3000, (c) 5.0 and 4000, and (d) 5.0 and 3000.

The resulting 36 histograms were subjected to a best-fit (chi-square) procedure, with scores that ranged from 0.001 to 0.009 and degrees of freedom that ranged from 4 (for higher-intensity samples) down to 1 (for low-intensity samples). None of the intensities thus tested were very high, as the concern was mainly about the relation between ε and r at low intensities. Such values are typical of the sample data that were used in the meta-study described in Chapter 8. In each case, the value for ε of the best fit was recorded. Figure 6.1 displays the results of this experiment as values of ε plotted against values of r. The results show high variability, as the best

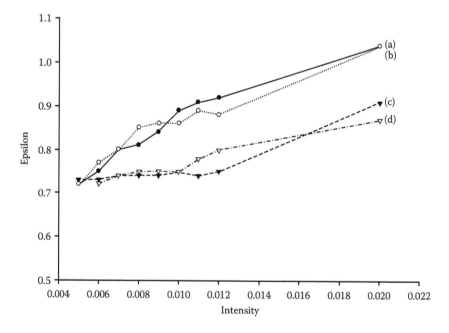

FIGURE 6.1 Plots of changes in sample ε values for increasing r and four combinations, (a–d) of ε and Δ.

fit response surface is rather shallow and the minimum value is easily influenced by slight changes in parameter values (see Section 4.4.1). But the basic trend is linear except near the low end, where intensities were below 1%. The extended straight line joining the low-intensity samples with the lone check sample is of course purely extrapolatory and not to be taken literally.

Before running these tests, I imagined that as the sample intensity declined, the trends in ε values would be downward and converging to 0.0. Instead, Figure 6.1 shows the trend line aiming at a point somewhat above 0.5 on the ε axis (to the left of the axis in Figure 6.1). This puzzled me, especially when a few additional tests revealed a stubborn unwillingness for ε' values to decline any lower than about 0.5. It occurred to me, however, that the minimum value of r should not be 0, but 1/N, in effect, as it made no sense to have a sample that was devoid of specimens. Thus, if x = 1 in the formula for the J distribution, we also have R = 1, Δ = 1, and F(1) = 1, which yields the following formula, after a bit of algebra:

$$1/(1+\varepsilon) = \ln (1/\varepsilon).$$

This equation has the unique solution $\varepsilon = 0.51734$, to five places of accuracy. This explains the mystery of the nonconvergent property of ε' as r approaches 0. This also explains why the two main trends appear to converge together as they approach the terminal low point.

6.2   ACCUMULATION CURVES

The accumulation curve is a theoretical shape that one has good reason to believe will be followed by an extended sample of a particular community. As more and more individuals are added to the sample, new species also appear, but at a declining rate. At first, they come in relatively often, but as the sample is extended, new species appear less and less frequently. The resulting curve, giving the number of species found so far as a function of the number of individuals sampled, appears to be logarithmic in shape, but is not necessarily. Sometimes called a rarefaction curve (owing to species becoming rarified as sampling proceeds), the accumulation curve comes in two distinct flavors, depending on whether the associated sampling process is carried out with or without replacement.

For example, if one is sampling without replacement and continues on to the bitter end, one runs out of individuals to sample. The sampling sequence is finite and the accumulation curve becomes flat. But if one is sampling a community with replacement, the sampling sequence is potentially infinite, as individuals already sampled may appear again (and again). In this case, the curve takes longer to flatten.

Although we shall display formulas for both kinds of sampling process, the formulas themselves are not simple, but involve summations. Moreover, they cannot be used in a field situation since abundances in the community must be known in advance of deployment. In both of the cases to be considered, F will represent the species abundance distribution of the community being sampled and R(n) will denote the accumulation function. After n drawings have been made, what is the expected number R(n) of species? Here, as elsewhere, N will represent the total population size of the community as a whole and R will represent the total number of species.

6.2.1 ACCUMULATION WITH REPLACEMENT

A formula developed in Appendix A.2.9 yields the exact *expected* number of species to show up after k samplings of the univoltine distribution. For the case discussed there, it is based on the recurrence relation,

$$R(k+1) = 0.9R(k)+1$$

This formula yields the k + 1st value of richness R in terms of the kth value. The solution of the recurrence relation ultimately yields a summation, as follows. The distribution that represents a community may be regarded as a serial collection of univoltine distributions, each representing a different abundance category. We merely apply that formula once for each abundance category, taking the context of N individuals into account.

Let $R_j(k)$ denote the number of species of abundance j (in the community) that are observed by the kth drawing. Considered in isolation, the number F(j) may be regarded as a univoltine distribution on its own and the total number of individuals in all species of abundance j yet to be sampled must be

$$j(F(j) - R_j(k)),$$

and the proportion of such individuals must be

$$(j/N)(F(j) - R_j(k)).$$

This is the probability that a new species will appear on the next drawing. Adding this number to Rj(k) yields the quantity on the right-hand side of the following equation, and the recurrence relation,

$$R_j(k+1) = (1 - j/N)R_j(k) + jF(j)/N.$$

Back-solving over several levels involves starting with some modest value for k such as 3, writing the equation for k = 2, then 1, and generalizing the resulting expression. The following formula emerges from this analysis:

$$R_j(k) = \frac{jF(j)}{N} \sum_{i=0}^{k-1} \left(1 - \frac{j}{N}\right)^i.$$

The full formula for R(k) consists of the terms Rj(k) simply added together, so that

$$R(k) = \sum_{j=1}^{\Delta} \sum_{i=0}^{k-1} m_j (1 - j/N)^i, \tag{6.2}$$

where $m_j = jF(j)/N$.

Note that the order of summation may be changed, if necessary, for convenience in calculating R(k). Also note that the inner summation is a finite power series in $x = (1 - j/N)$. As shown in Appendix A.2.2, the corresponding infinite series sums to $1/(1 - x)$, which becomes simply N/j in cases where x is small enough to converge quickly. In order to split the sum of Equation 6.2 into convergent and nonconvergent parts, a convergence approximation may be used. How large must k be, for example, in order that the inequality, $x^k \le 0.01$ be satisfied? After a bit of algebra, one obtains

$$j \ge N(1 - (0.01)^{1/k}).$$

Denoting the quantity on the right-hand side of the inequality by k', we rewrite Equation 6.2 as

$$R(k) = \sum_{j=1}^{k'} m_j \sum_{i=0}^{k-1} (1 - j/N)^i + \sum_{j=k'+1}^{\Delta} mj \sum_{i=0}^{k-1} (1 - j/N)^i$$

Since $\sum_{i=0}^{k-1} (1 - j/N)^i \approx N/j$ in the second term and $mj = jF(j)/N$ in both terms, we have

$$R(k) \approx \sum_{j=1}^{k'} \frac{jF(j)}{N} \sum_{i=0}^{k-1} \left(1 - \frac{j}{N}\right)^i + \sum_{j=k'+1}^{\Delta} \frac{jF(j)}{N} \left(\frac{N}{j}\right)$$

$$= \left(\frac{1}{N}\right) \sum_{j=1}^{k'-1} jF(j) \sum_{i=0}^{k-1} \left(1 - \frac{j}{N}\right)^i + \sum_{j=k'}^{\Delta} F(j). \tag{6.3}$$

In other words, all species with abundances of k' or greater are virtually certain to contribute to the richness of the sample by the kth observation in the community. Contributions by species of lower abundance are conditioned by other factors in the first summation. As k becomes larger, the quantity k' becomes smaller and the second summation increasingly dominates R(k). It is not clear whether the formula as a whole can be encapsulated by a simple function of logarithmic or any other form. Unless one is adroit with a calculator, the simplest way to use Equation 6.3 is to employ a program for the purpose.

TABLE 6.1

Comparison of Theoretical (R) and Sample (S) Richness Estimates

k	5	10	15	20	25
r	0.1	0.2	0.3	0.4	0.5
S	4.25	7.15	8.50	10.25	11.20
R(k)	4.21	6.96	8.84	10.19	11.19
Error	0.04	0.19	0.34	0.06	0.01

The accuracy of the accumulation formula embodied by Equation 6.3 is illustrated by a very simple distribution consisting of 10 species of abundance 2 and five species of abundance 6. A program called Accum was written to perform the calculation of R(k) for any value of k. Table 6.1 summarizes the results when the values of R(k) are compared with the corresponding output S of the program SampleSim, with intensity r varying from 0.1 to 0.4 and 10 trials per case.

The relative errors in this example begin at 0.95% and decline to 0.09% in the last case. The formula for R(k) is not of much practical use because one has to know the values F(j) of the community distribution in order to use it. However, it shows clearly how the number of species expected to show up by the kth sampling step depends on F. This amounts to yet another demonstration of the importance of knowing the distribution that prevails in the community.

6.2.2 ACCUMULATION WITHOUT REPLACEMENT

6.2.2.1 Hurlburt's Formula

In this section, we present an early example of exact methods in the development of accumulation curves, followed by a new formula based on the canonical sequence of the J distribution.

The without-replacement accumulation curve found by Hurlbert (1971) relies on knowledge of abundances a_i in the target community and relies heavily on the combinatorial choice function represented here by the notation

$$C(n,k) = n!/(n-k)!k!.$$

The function C denotes the number of ways of choosing k things from among n. Equation 6.4 employs the notation a_i to represent the abundance of the ith species in the community, the exact order of species being arbitrary but fixed. As in the previous section, the function R(n) will denote the number of species to have appeared by the nth observation.

$$R(n) = R - C(N,n)^{-1} \sum_{i=1}^{R} C\big((N - a_i), n\big) \tag{6.4}$$

A brief demonstration of the accuracy of this formula is provided by the univoltine distribution, c U[10] × 20. Here, N = 200, R = 20, and a_i = 10 for all values of i. Equation 6.4 becomes

$$R(n) = 20 - C(200,n)^{-1} \sum_{i=1}^{20} C(190,n)$$
$$= 20[1 - C(200,n)^{-1} C(190,n)]$$

Using a hand calculator, it will absorb many busy minutes to discover that

$$R(1) = 1.0$$

and that

$$R(5) = 4.66.$$

Running the program SampleSim (see Appendix A.3.2) on the distribution U[10] × 20 at intensity r = 0.025 (sample size 5), we find an average value for R(5) over 20 samples to be 4.64, about as close as one can reasonably expect.

It must be remarked that neither Hurlburt's accumulation formula nor the one that appears in the previous section seem capable of further simplification. Consequently, it is difficult to determine easily the shape that either curve must have.

6.2.2.2 The Hyperbolic Formula

An accumulation curve that can be expressed in a compact formula is readily obtained from the J distribution (for the community in question) by inverting the density function: namely, exchanging the dependent and independent variables in effect, as described in Section 1.1. Treated in this manner (turning the density function on its side, so to speak), an accumulation formula emerges from the inverted density function in the case of sampling without replacement. Figure 6.2 displays the density function in

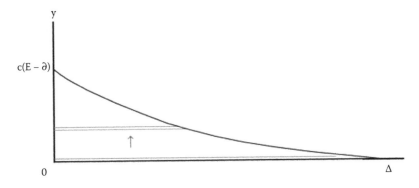

FIGURE 6.2 Integrating the inverted density function.

its normal (noninverted) position, with two infinitesimal horizontal strips that represent elements of integration acting on the variable y—as if the figure were turned on its side.

The density function of the J distribution,

$$y = c(1/(x+\varepsilon)-\delta),$$

is readily inverted to yield an equivalent formula in which the roles of dependent and independent variable have been interchanged.

$$x = c/(y+c\delta)-\varepsilon$$

The basal strip in Figure 6.2 represents an element of integration that may be thought of as a miniature replicate of the species densities as manifested in the categories of lowest abundance to those of highest abundance. In other words, viewed as a probability distribution in which the tiniest widths are allowed to the species of abundance 1, right out to the greatest widths that correspond to all the population sizes that a species of greatest abundance might have, each category within the strip occupies a length that is proportional to its abundance. A single individual is drawn according to these probabilities. Higher up, a second strip illustrates the process of repeated sampling at a more advanced stage, say at the kth drawing. This strip is shorter, as each category within it is now shortened by the number of individuals that have already been selected for the sample.

The integral that corresponds to this advanced stage of the sample may be defined as follows,

$$A'(k) = \int_0^k \left(\frac{c}{y+c\delta} - \varepsilon \right) dy$$

and integration results in the formula,

$$A'(k) = c\ln(y+c\delta) - c\varepsilon y \Big|_0^k$$
$$= c(\ln((k+c\delta)/c\delta) - c\varepsilon k$$
$$= c(\ln(k/c\delta+1) - \varepsilon k)$$

The variable y ranges from 0 to c(E − δ), the maximum value that y can have. (Here, E represents the inverse of ε.) To use this formula effectively in the context of the hyperbolic theory, it will be necessary to convert the independent variable from y to r, the intensity of the sample (as extended so far) in relation to the community. The resulting formula, denoted by A, replaces the variable y by c(E − δ)r and applies the resulting ratio to R, the richness of the community being sampled. The following formula results:

$$A(r) = Rc[\ln((E/\delta-1)r+1)-(1-\delta)r] \tag{6.5}$$

Worked example: Let C = J(10, 1000) × 500, noting that the parameter delta is given in the Δ-form. In this case, the formula becomes

$$A(r) = 500(0.277)(\ln((99)r + 1) - (1 - 0.0001)r)$$
$$\approx 138.5(\ln(99r + 1) - r),$$

and the following calculations hint at the logarithmic shape of the curve. The −1 term at the end of each calculation reflects the fact that in all cases, $(1 - 0.0001)r$ is very close to unity.

$$A(0.001) = (138.5)(0.094) - 1 = 12.9$$
$$A(0.010) = (138.5)(0.688) - 1 = 93.9$$
$$A(0.100) = (138.5)(2.389) - 1 = 317.0$$
$$A(1.000) = (138.5)(3.605) - 1 = 499.3$$

Are these numbers realistic in relation to actual field samples? Since the formula is based on the J distribution it comes as no surprise that A(1.0) = 499.3. At a relative sample size of 0.001, however, 12.9 seems to be a lot of species until one realizes that C has a total population of N = 62,879 individuals and that r = 0.001 amounts to a sample size of 63 individuals. In any case, the development just given cannot be used in a field situation unless one already knows the J-parameters of the community (see Chapter 5).

6.3 THE SPECIES–AREA RELATIONSHIP

Consider a region H of uniform habitat, an area A within H, and a community C that has the distribution F = J[ε, δ] × R, where R is the number of species in C. Suppose now that the Area A is expanded to include an additional area a, the amalgamated area becoming A + a.

Assuming spatial uniformity of A, it is reasonable to assume that, on average, the abundance of each species in the newly enlarged community has increased by the proportion,

$$\gamma = a/A.$$

In deriving new abundances, however, it is not clear where one takes the zero-point as the basis of multiplications. For example, if we take the zero point at the origin, a kind of tautology results: A species of abundance k would end with an expected abundance of $k(1 + \gamma)$ following the amalgamation. If we apply the transformation $x \longrightarrow (1 + \gamma)x$ to the distribution F, we obtain the new distribution F' (after some algebra):

$$F'(x) = Rc(1 + \gamma)(1/(x + \varepsilon') - \delta'),$$

where $\varepsilon' = (1 + \gamma)\varepsilon$ and $\delta' = \delta/(1 + \gamma)$.

This implies that c', the new coefficient, is identical to c. Moreover, we have forgotten that, from the point of view of a continuous distribution, the gaps left by the transformation applied to the discrete form must show up in the continuous form following the division $1/(1 + \gamma)$. This results in the simpler expression, but the same number R of species:

$$F'(x) = Rc(1/(x+\varepsilon')-\delta').$$

The only alternative approach involves using the zero point of the untranslated hyperbola, as shown in Figure 6.3. When the same transformation $x \longrightarrow (1 + \gamma)x$ is applied to the zero-point of G, it merely has to be asked how many "species" implied by the distribution G (i.e., from the ε-zone shown in Figure 6.3 below) end up at or beyond the zero point of F.

When the transformation is applied to G, all abundances in the G interval $[\varepsilon/(1 + \gamma),$ ε] now appear in the transformed version F' of F. The resulting distribution may be written J[ε', δ'], where

$$\varepsilon' = \varepsilon / (1+\gamma) \quad \text{and} \quad \delta' = \delta/(1+\gamma).$$

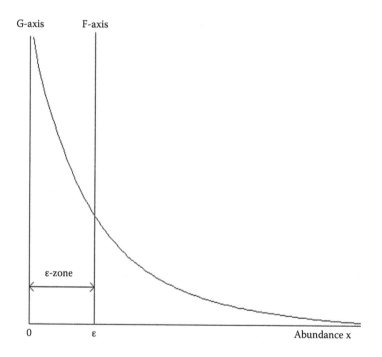

FIGURE 6.3 The ε-zone considered as the source of new species.

The second equality is approximate but plays no further role in this analysis in any case. In terms of the given parameters ε and δ, the new c value becomes

$$c' = \left((1+\gamma)^2 \ln(\Delta/\varepsilon) - 1\right)^{-1},$$

which is clearly smaller than c. The number s of new species would, by this method of determination, be given by the following indefinite integral of the untranslated function G,

$$s = Rc \int G(1/x - \delta) dx,$$

taken over the interval $[\varepsilon/(1 + \gamma), \varepsilon]$. The corresponding definite integral may be simplified by neglecting the δ term entirely, yielding the integral

$$s = Rc \int_{\frac{\varepsilon}{(1+\gamma)}}^{\varepsilon} \frac{dx}{x} \tag{6.6}$$

$$= Rc \, (\ln(1+\gamma))$$

It may be noted that the vertical line that marks the lower limit of abundance in the original sample resembles the veil line (Preston 1948) but differs in three key respects: First, species "cross" the line in the opposite direction. Second, the operation just described is not an additive transformation, but a multiplicative one. Finally (and most importantly), it is not applied to a sample, but to a whole community.

Worked example: Let C have the distribution J[5.0, 0.005] × 90 within an area A of 100 ha. If another 30 ha is added to area A for a total of 130 ha, the value of $1 + \gamma$ becomes 1.30, which yields $\ln(1 + \gamma) = 0.262$. Since R is 90 and c turns out to be 0.501, we end with

$$s = (90)(0.501)(0.262) = 11.83.$$

On average, about 12 new species will be included following the annexation. A higher value of ε (δ being fixed for the moment) would result in a lower value of c and, therefore, a smaller number of new species.

6.3.1   OTHER SPECIES–AREA LAWS

The species–area relationship most commonly described in the literature takes the form of a power law, for example, "...the number of species is proportional to the size of the area in which they are found, raised to an exponent d (usually a number between 0.2 and 0.3)" (May and Stumpf 2000)

$$R \alpha A^d$$

or

$$R = bA^d$$

where b is an appropriate constant.

In order to bring such a formulation into harmony with the incremental approach used here, we increase the area A by the amount a (as previously) to obtain the number s of new species found when the area is increased by the proportion a/A (= γ)

$$
\begin{aligned}
s &= (R+s) - R \\
&= b((A+a)^d - A^d) \\
&= bA^d((1+\gamma)^d - 1) \\
&= R((1+\gamma)^d - 1)
\end{aligned}
$$

The factor $(1 + \gamma)^d$ clearly resembles the factor $\ln(1 + \gamma)$, given the small size of the exponent d, and the two expressions may yield values that, in some cases, would be close enough to escape notice in the field, so to speak. However, they give rise to rather different results than the logarithmic formula (Equation 6.5), in general. Taking derivatives of the respective functions with respect to γ reveals a marked difference in their respective behaviors:

Logarithmic formula: $D\gamma = (1 + \gamma)^{-1}$
Exponential formula: $D\gamma = c(1 + \gamma)^{1-c}$, where $1 - c > 0$

According to the respective derivatives, the slope of the first function, although always positive, decreases modestly with increasing γ, while the slope of the exponential expression increases without limit. Clearly, the two formulas cannot both be correct. In any case, the exponential formula cannot be considered a part of the hyperbolic theory and is unlikely to play any role in it.

The logarithmic formula (Equation 6.5) can be applied with statistical precision only if the parameters ε and δ are known for the community in question—or if one has good estimates for them, as determined, for example, by the methods of Chapter 5. Since different communities within area A will, in general, have different values of these parameters, one cannot extend the law automatically to the overall species richness of the area since the actual multiplier Rc will be different for each community C and the proportionality constants will therefore all be different. However, if one took all the species living in the area A as the "community" C and one had a valid estimate for the overall values of ε and δ, the more general law could nevertheless be applied.

7 Stochastic Systems and the Stochastic Community

The stochastic species hypothesis and its variants form the centerpiece of this chapter. Central to the hypothesis is the notion of innate probabilities that cannot be observed or measured directly. The same notion is central to the theory of evolution (probability of survival), so it is not a new *kind* of concept.

Interestingly, the same notion also lies at the heart of the most successful physical theory ever discovered—quantum mechanics (Herbert 1985). Photons that pass through a single slit may appear anywhere on a screen according to the (statistical) dictates of the wave function, a "probability wave" that is not considered to have any physical reality until it is measured. As such, it merely governs the overall distribution of photons on the screen (Rae 1986, p. 10). The reality of the underlying probabilities is inferred from the reality of the pattern so manifested.

The conversion of a photon in its wave (probability) form to a small flash of light on the screen, its particle (manifest) form, involves the infamous "collapse of the wave function," a seemingly random point on the screen where the photon manifests and becomes visible. The placement of all such points nevertheless follows the Airy distribution, as shown in Figure 7.1. The precision with which the manifesting pattern of flashes on the screen "fill out" the Airy distribution cannot be gainsaid; the more photons that arrive at the screen, the more fully and accurately that distribution (and no other) appears. Such probabilities can only be "measured" after the fact. Similarly, it is only by counting the births and deaths within a population over time can we infer anything about the hidden probabilities. And of course, we may get it wrong. The births and deaths manifest such probabilities only in the long run, as the expression goes. I would not claim the same *degree* of precision for the theory developed in this monograph, merely the same *kind* of precision.

7.1 THE STOCHASTIC COMMUNITY AND STOCHASTIC SYSTEMS

Depending on the context, the "stochastic community" referred to in the title of this chapter refers to any natural community that has the J-attribute, thus *appearing* to follow the J distribution, a density function that is populated by R species, as follows:

$$F(x) = Rc(1 / (x + \varepsilon) - \delta).$$

The remainder of this chapter is devoted to exploring the dynamic behavior associated with the stochastic species hypothesis as manifested by a class of stochastic

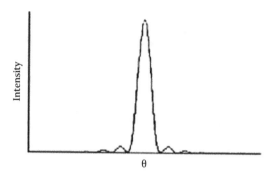

FIGURE 7.1   The Airy pattern.

systems and, to some extent, by natural communities, as well. Viewing a natural community as a dynamic system, we may readily imagine populations fluctuating over time, some increasing, some decreasing, others retracing their steps. A myriad of physical events and natural interactions influence the ups and downs of every population in a manner that (with a few caveats) appears to be effectively random, as described earlier in Section 3.1. In this chapter, I describe a few abstract systems that manifest the stochastic species hypothesis.

The behavior of such systems resemble random walks, with surprises in store for those who may not have realized the full range of behaviors of which such excursions are capable. This may help to account for (if not explain) the mysterious persistence of certain species.

7.1.1   PROBABILITY AND TIME

If we observe a population over time, births and deaths would appear to occur at random and a time series may be constructed, as illustrated in Figure 7.2.

The "population" in question may arise from a stochastic system or, by way of hypothesis, from a natural community of the kind discussed in this monograph. In either case, such a time series is conventionally described by the negative exponential distribution, as follows (Feller 1968, Thornton 2013):

$$f(t) = (1/\lambda)e^{-t/\lambda}; \ t > 0. \ \text{(The formula is proved in Appendix A.2.6.)}$$

The density function gives the distribution of interevent times, of which the average is the wavelength λ. For large values of t, the function is very small, meaning that the

t 0 3   8 9   20...   ...181

FIGURE 7.2   Birth and death events form a random time series.

occurrence of interevent times of such magnitude happens only rarely. Such a time series may be used to describe either birth times or death times.

The formula allows one to be more precise about the application of probabilities that lie at the heart of the stochastic species hypothesis and all developments that arise from it. Suppose that the previous time series represents births and let p_t represent the probability of a birth within a time period of duration t. The exponential distribution yields an exact value for p_t. It is simply unity minus the relative f-density of all those instances where the interevent interval exceeds the time unit t.

$$
\begin{aligned}
p_t &= 1 - \int_t^\infty (1/\lambda)e^{-t/\lambda}\, dt \\
&= 1 - (1 - e^{-t/\lambda})\Big|_t^\infty \\
&= 1 - e^{-t/\lambda}
\end{aligned}
\tag{7.1}
$$

This relationship between t and λ is invertible and one has (writing λ as a function of t)

$$
\lambda(t) = -t/\ln(1 - p_t).
$$

The quantity so derived is positive in spite of appearances to the contrary; the natural logarithm of a number less than unity is itself negative. A similar relationship involving the wavelength associated with the probability q_t of deaths may also be derived. If q_t differs from p_t, the two wavelengths must also differ. The wavelength formula will be used in the derivation of a variant of the stochastic hypothesis that applies to populations over extended periods of time.

7.1.2 GENERALIZATIONS OF THE MULTISPECIES LOGISTIC SYSTEM

The basic multispecies logistic system (MLS), as described in Section 1.2, is capable of almost endless generalizations, being what a systems developer would call "detail hungry." The main generalization, as defined immediately in the following, gives rise to several variants, shortly to be described. In the present context, the variants are all dynamical systems, with no immediate application to natural communities in this section.

A *stochastic system* is a set C of N elements (called *individuals*) that is partitioned into R nonempty subsets (called *species* or *populations*, according to context). The species are indexed by the integers, and S_1, S_2, S_3,..., S_R will represent the size of their respective populations, so that

$$
\sum_{i=1}^R S_i = N.
$$

The system obeys dynamic rules that incorporate a clock based on a fundamental time unit, τ. At each tick of the clock (passage of one time unit), an individual is selected at random from C and is either duplicated (birth), deleted (death), or left alone, resulting in a new value for the population S_i to which it belongs, with an associated adjustment in the value of N. The decision to increase or decrease the population S_i is made on the basis of equal probabilities. Stochastic systems embody the stochastic species hypothesis.

The foregoing definition automatically implies a time series of the kind described in the previous section. A species of abundance k will contain k individuals, and one of these will be chosen, on average, with a frequency proportional to k/N. It follows that the average time between events (or wavelength) is N/k, so that in Equation 7.1, λ may be set to N/k, resulting in an expression for p_t,

$$p_t = 1 - e^{-kt/N}. \tag{7.2}$$

Although the probability of change affecting individuals does not itself change, the probability affecting whole species (i.e., populations) certainly does. Obviously, the larger the population, the more frequently it changes its abundance. It is highly remarkable (and surprising to those unfamiliar with such systems) that a stochastic system does not produce what ecologists call "regulated" populations. Quite the contrary, a large number of populations will drift toward low abundance, while a few populations will drift, with equal certainty, to high abundances. As the system progresses, some of the low-abundance populations become numerous again, while one or more of the high-abundance populations will decline in number. All of this happens without the two probabilities ever failing to be equal.

Example: The MSL system (see Section 1.2) is obviously a stochastic system in which $p_i = 0.5$ for all index values i = 1, 2, 3,..., R.

The following variation on stochastic systems is included to illustrate the open-ended nature of the kind of systems that can be defined under this heading. If the probabilities defined for a stochastic system are redefined as in Table 7.1, the result will be called a *weakly stochastic system*: the probabilities p_i and q_i are themselves allowed to vary in a manner that implies a long-term equality. For example, the probabilities p_i might be normally (or binomially) distributed, as in Table 7.1.

At each iteration of the system, an individual is selected at random from the community and the nature of the event is decided by first selecting a row in the "distribution" column according to the probabilities that appear there. A random number drawn from the interval [0.0, 1.0] will determine which row of the table will become operative. The central row will be selected with probability 0.3125, either of the flanking rows with probability 0.2344, and so on. Next, the corresponding probability p_i is determined by a second random number drawn from the interval [0.0, 1.0]. If the number falls within the interval $[p_i - 1, p_i]$ the individual reproduces.

Example: The StoComm program (see Appendix A.3.1) uses Table 7.1 to govern probabilities of births and deaths. Figure 7.3 shows a histogram of the distribution produced by such a system. The histogram records the long-term (average) behavior of the number of species in each abundance category.

TABLE 7.1

Table of (Birth) Probabilities

Probability, p_i	Distribution
0.1	0.0000
0.2	0.0156
0.3	0.0938
0.4	0.2344
0.5	0.3125
0.6	0.2344
0.7	0.0938
0.8	0.0156
0.9	0.0000

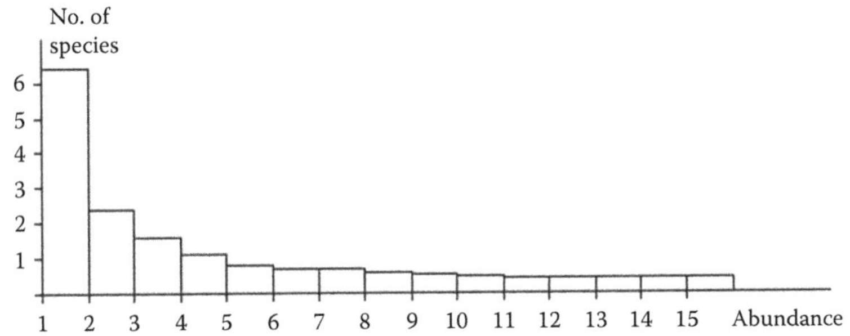

FIGURE 7.3   Average distribution of abundances produced by the StoComm program.

In this case, StoComm was inhabited by 50 species that underwent 50,000 itera-tions of the basic procedure. The chi-square score for this histogram, when com-pared to the J distribution, is 2.489 at 5 degrees of freedom. This is a relatively close fit (5.000 being the average under the null hypothesis) but serves as little more than an indication of the presence of the J distribution. Such a run would have to be repeated many times; a continuing decline in chi-square scores toward zero would imply the J distribution more strongly.

It cannot be ruled out, in the present state of development of the hyperbolic the-ory, that weakly stochastic systems are merely stochastic systems in disguise, so to speak, being mathematically equivalent to them.

7.1.3   STOCHASTIC ABUNDANCES

If the J distribution describes abundances in communities of species, the normal distribution describes *changes* in those species. This could be stated as a theorem, but it amounts to little more than a simple deduction. Imagine the random walk on the abundance axis that would be performed by a stochastic species. One could call such peregrinations stochastic vibrations; now the population gains one, later it loses one. Sometimes, it gains (or loses) several individuals in a row. Since the probabilities for both birth and death hover about 0.5, population changes over a fixed number k of such events will follow the binomial distribution according to the Central Limit Theorem (Feller 1968). This well-known result states that if one carries out the following experiment many times on an arbitrary distribution F, the distribution of the random variable G will tend to be normal:

Let $X_1, X_2,..., X_m$ be m random drawings from F and let G be the random variable defined as follows:

$$G = (X_1 + X_2 + X_3 + ... X_m)/m.$$

The more observations of G that can be compiled, the more closely the histogram formed by such observations will approximate the binomial distribution, which, in turn, approximates the normal distribution. Moreover, the mean μ of G will approximate the mean of F. In the case of a stochastically vibrating population, the changes that occur over fixed periods of k events (births and deaths) tend to be evenly balanced, producing an expected mean value of zero, since for every increase of x individuals, there tends to be a corresponding decrease.

Theorem 7.1: In a stochastic system, the expected change in the abundance of a species over time follows the binomial distribution with mean zero.

Given the foregoing reference to a more general theorem, no proof is needed for this one, but the details are nevertheless instructive in the context of this chapter.

Proof: Let s be a species for which the average time between births is λ. Then its birth probability p_t over a period of t time units will be given by Equation 7.2. Now consider m consecutive time periods over which the probability may apply. This yields the recipe for a binomial distribution $B(p_t, m)$ with density values given by the formula

$$B(k) = C(k,m)p_t^k,$$

where $0 \le k \le m$. A similar formula holds for death probabilities. If one populates this formula by the births that occur within each period of m time units, one obtains the corresponding histogram, which follows the binomial distribution function. Therefore, denote by the series $b_1, b_2, b_3,...$ the number of births that occur in consecutive time periods and by $d_1, d_2, d_3,...$ the corresponding number of deaths. If both sets of random variables were distributed normally, so would be the new random variables $b_i + d_i$ and $b_i - d_i$ (Neuts 1973). This behavior is illustrated dramatically by the well-known Galton board, as in Figure 7.4.

In a Galton board (Wolfram 2016), balls are introduced one at a time at the top of the device. Falling downward, each ball strikes a succession of k pegs, bouncing to the left (death) or to the right (birth) with equal probability. After k bounces, most of the balls end up close to the average at the center, but many other balls end up further away.

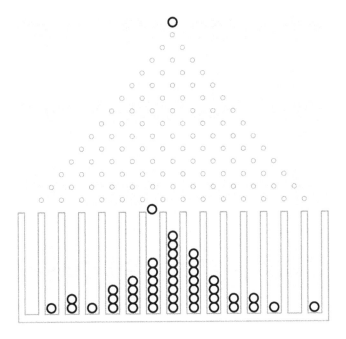

FIGURE 7.4   Balls representing population changes produce a binomial distribution.

To apply the Central Limit Theorem in the presence of the stochastic species hypothesis, one merely recognizes that changes in a population, as observed periodically, amount to sums of k small changes that make up the counts. As such, the sums amount to averages, even though averages are not taken. If one actually took averages, dividing by k, all the numbers would be smaller, but the resulting histogram would simply be a scaled-down version of the distribution of counts.   ■

7.2   STOCHASTIC COMMUNITIES IN NATURE

In the foregoing section, it was shown that the stochastic species hypothesis implied an equiprobable change of abundance in either direction, resulting in a binomial distribution of changes in abundance at the species level. This section explores the presence of such distributions in natural data in support of the stochastic species hypothesis. This results in a test that can be applied to single populations in nature, communities aside.

7.2.1   STOCHASTIC SPECIES HYPOTHESES FOR COMMUNITIES

It is doubtful that one could argue, purely on the basis of some well-known collective property, that natural communities operate under the stochastic species hypothesis. On the other hand, one may adopt the property of Theorem 7.1 as a companion hypothesis and then derive the stochastic species hypothesis from it. Let us imagine, then, a real population in which the expected changes of abundance over time follow the binomial distribution. To make the analogy with the stochastic systems discussed in the foregoing section a little stronger, we may also imagine a tropical setting with no major disturbances and the ability of its species to breed (or die) as a year-round possibility. Suppose that S is a particular species in the community for which (thanks to some brilliant field work) we have a complete time series of reproduction events. On such a basis, one can estimate the probability p empirically that an individual of S will reproduce. We add the somewhat unrealistic assumption that just one off-spring results from breeding and that no males are counted in the population to be analyzed. In this analysis time will be measured in days.

Let the total number of births over a year be a and let the total number of deaths be b. If species S has a population of n at the beginning of the time series, it will have $(n + a - b)$ at the end of it. The average time between births will be λ_a (as in Section 7.1.1), for which the fraction 365/a provides an (temporary) estimate in days-as-wavelength. Given Equation 7.1, we may write a formal expression for the probability that the species in question will reproduce over time t. This expression will act as a theoretical bridge to the stochastic species hypothesis, given the binomial distribution of population changes.

$$p_a(t) \approx 1 - e^{-t/\lambda_\alpha}$$
$$= 1 - e^{-at/365}. \tag{7.3}$$

Similarly,

$$p_b(t) \approx 1 - e^{-bt/365}, \tag{7.4}$$

where p_b is the corresponding probability of a death.

In the same sense that an expression like 365/a forms an estimate for λ_a, the quantities $p_a(t)$ and $p_b(t)$ form estimates of the respective probabilities, as indicated by the $\approx$ symbol. If we compare the expressions in Equations 7.3 and 7.4 in the form of a ratio, the resulting expression becomes

$$e^{-at/365}/e^{-bt/365}$$
$$= e^{(b-a)t/365}. \tag{7.5}$$

If the foregoing observations of a and b were to be performed many times, a variety of values for the latter ratio would result. Some of these might have a negative exponent, and some, a positive one. The differences of two random variables that follow the binomial distribution also follow the binomial distribution (Neuts 1973). Thus the differences $(b - a)$ follow the binomial distribution. In the long run, the mean will tend to zero since every birth must sooner or later be accompanied by a death. The average value of the Equation 7.5 would therefore tend to unity. This would imply not only the ultimate equality of the two exponential expressions but also the equality of $p_a(t)$ and $p_b(t)$.

The foregoing community was placed in an equable tropical setting. What about temperate communities in which seasonal effects have a strong influence on births and deaths? Here, the argument is more difficult to make, with only one simple observation to support it. The birth and death time series for a temperate zone population of mammals, for example, would consist (in the spring) of a sequence of birth events with a death event or two interposed. This would be followed by nothing but death events strung out over the summer, into the fall, and through the winter, where mortality would probably increase. A histogram of population changes over the year would consist of a high bar or two on the positive side of zero (the births) and something resembling the negative half of a binomial distribution on the negative side of zero. If it could be shown that the mean of such distributions still tended to zero, the previous argument could be extended to temperate zone communities as well. However, the difficulty raised here would not necessarily apply to the year-over-year statistics, as these would each amount to a single number only, the difference between last year's population and the present one.

7.2.2   DETECTING THE J DISTRIBUTION IN NATURAL POPULATIONS

If it could be shown that the annual fluctuations of natural populations follow the binomial distribution, one would have strong evidence for the presence not only of the stochastic species hypothesis but also for the J distribution in communities, automatically. In this section, we simulate such fluctuations and compare them with a few datasets from real populations, paving the way for additional research to be described in Chapter 10. The aim of this mini-experiment was to evaluate the disorder or "raggedness" of the resulting histograms. The simulated population data were obtained from a run of the MSL program in which one species was recorded at every 100 iterations of the basic algorithm, each comprising one "interval." Although we expect them to follow the binomial distribution, the degree of closeness is statistical in nature, as can be seen in Figure 7.5a and b.

If it were not known that these histograms arose from the binomial distribution, one might not suspect it, especially of the smaller chart. In fact, the smaller histogram represents the data after 15 counts, with another 48 population counts resulting in the larger one. However, the larger histogram in Figure 7.5b is certainly closer to the binomial shape than the earlier version. Were one to continue generating counts from the MSL system, the shape would gradually approach that of the binomial (or normal) and no other.

Turning now to natural populations, histograms of three population counts appear in Figure 7.6a, b, and c. These represent annual counts of three animal populations, sea otters (Hatfield and Tinker 2013), brook trout (Waters 1999), and grizzly bears (Morris et al. 1999).

In these histograms, data have been grouped to bring out the essential shape of the distribution of abundances. In Figure 7.5a, the multiple is 100; in Figure 7.5b, the multiple is 1000; and in Figure 7.5c, the multiple is 2. Since each category spans a range of abundances, upper limits (times the multiple) label the right-hand end point of the respective intervals.

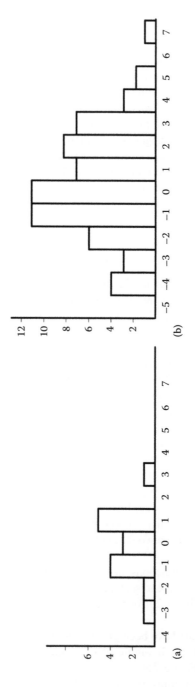

FIGURE 7.5 Appearance of the histogram after 16 (a) and 63 (b) intervals, respectively.

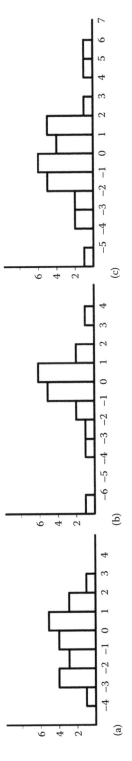

FIGURE 7.6 Counts of (a) sea otters, (b) brook trout, and (c) grizzly bears.

The resemblance to the binomial distribution among the latter histograms is certainly no worse than the resemblance of the corresponding simulated histogram in Figure 7.4a. In fact, Figure 7.6a is slightly closer to a binomial shape than the histograms of Figure 7.5. A better example was obtained from the Breeding Bird Survey data (Robbins et al. 1986), a massive database of some 60,000 site surveys taken over the years across North America. The average of histograms of variability in the Rose-breasted Grosbeak from 100 sites chosen at random was plotted in this manner. The resulting shape is shown in Figure 7.7.

Here, we see a strong central tendency at zero fluctuations. The shape of the histogram is naturally ragged. According to the stochastic species hypothesis, as described in Section 7.1.1, the shape should be a binomial distribution tending to normality. As far as establishing a binomial shape in annual census data (such as the example used earlier), a more thorough investigation would proceed by producing a succession of ever-larger averages in order to discover whether the shape converges to an even more definite normal shape and how quickly it does so.

7.2.3 THE STOCHASTIC ORBIT AND ITS VARIATIONS

Besides the time series for a species, the *stochastic orbit* is another important conceptual tool. A species does not progress evenly through its abundances (as a planet progresses evenly through its positions), but erratically, rather in the manner of a random walk, with a high frequency of "vibrations" at high abundances and a low frequency at low abundances. At the high end, total biomass/energy conservation begins to constrain the orbit statistically, while species at the low abundance end have slowed to a relative crawl. Indeed, over time (perhaps an extremely long time) and under suitable measurement constraints, a single species may recapture the J distribution all on its own. The probability of finding a given number of species at a given position is governed by the J distribution. It may even be the case that a community of species with very different lengths of reproductive cycle will still produce a J-shaped distribution. They would manifest in a manner similar to that of photons in the Airy distribution, as explained at the beginning of this chapter.

This view of how the abundances of species behave over time stands in distinct contrast to more traditional views that assumed—and then sought evidence for—stability in populations. The search for stability has not been a great success. (See the quote from Daniel Botkin in Section 1.6.) In his Robert H. MacArthur Award Lecture, William Murdoch (Strong 1991) illustrated the difficulty of establishing the presence of density dependence in various populations under study. The notion of "regulated populations" was slowly giving way to nonequilibrium population dynamics, an approach that began in the late 1960s (Den Boer 1968, Levins 1969) and developed subsequently in Chernov et al. (1980), Koetsier et al. (1990), and Murdoch (1994). In the nonequilibrium dynamical view, local populations are free to fluctuate randomly, with local extinctions more or less balanced by immigration from neighboring communities. Indeed, the J distribution may be regarded as a strong candidate for the "fixed probability distribution" proposed by Murdoch. The intermediate notion of "stochastic boundedness" (Chesson 1978, for example) sought to harmonize the two extremes by invoking density dependence as a mechanism

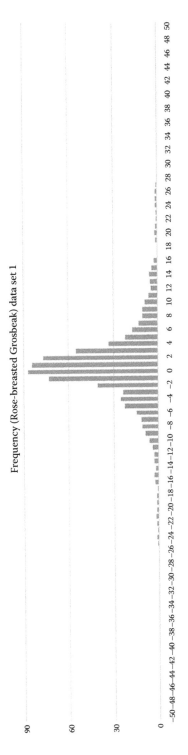

FIGURE 7.7 Fluctuations in annual counts of the Rose-breasted Grosbeak.

only at the extremes of high or low population densities. But even this faint hope was undermined by the observations of Den Boer (1990). "It was found that boundaries (Log Range) between which numbers fluctuate in field populations increase with time to about the same extent as in comparable random walks in density."

In the same paper, Den Boer points out, "Although it is not very probable that field populations fluctuate exactly like random walks of densities, random walk models appear to mimic the fluctuation patterns of field populations sufficiently closely to explain what happens in nature, and to deny the need for regulation." The emerging consensus on the possibility that populations are "regulated" in some manner was captured by Wolda (1995) two decades ago. "The quest for answers to the density dependent regulation controversy has continued over several decades now, and not only haven't we found satisfactory answers, we haven't even been able to point to a generally acceptable way of finding those answers."

In commenting on environmental stochasticity, Lande et al. (2003) put the matter more precisely, "Fluctuations in population size often appear to be stochastic or random in time, reflecting our ignorance about the detailed causes of individual mortality, reproduction and dispersal. The description and prediction of stochastic population dynamics in time and space is fundamental to ecology and conservation biology."

Suffice it to say that the most recent approach to population dynamics is entirely consistent with the theory proposed in this book. One might add the caveat that the search for equilibrium appears to be misdirected when aimed at single populations and not whole communities, where stable equilbria actually exist in a collective sense.

7.2.4 LONG RUNS IN STOCHASTIC BEHAVIOR

One of the seeming bars to general acceptance of random walks lurking in our field data are the twin notions of dominance and succession. The idea of species being dominant involves a population that is significantly and persistently larger than those of a group of species with which it is being compared. As will presently be made clear, perhaps to the surprise of some readers, lengthy stretches of dominance patterns over time are accommodated by the stochastic behavior that I have called "effective randomness" in Section 1.5.

In some plant communities, whether of woody or herbaceous type, the well-known phenomenon of succession involves a gradual change in the composition of a plant community, usually following a disturbance (Ricklefs 1990, p. 677). For example, in a specific area of Indiana, United States, the woody vegetation that was present 120,000 years ago was erased by the glaciers of the Wisconsinan Period (110,000 BP to 10,000 BP). Post-Wisconsinan, a large area of mid-continent North America, for example, followed a slow succession as woody species invaded from the ice-free lands to the south. First came the conifers, then the mixed hardwood/conifer regime, then the maple-basswood forest, followed by the beech-maple type (Braun 1972). Given the relatively stable climate over the area in question during the last few thousand years, beech-maple forests remained, for the most part, as beech-maple forest, with American beech (*Fagus grandifolia*) and sugar maple (*Acer Saccharum*) remaining as codominants in the canopy. Table 7.2 shows the relative

TABLE 7.2

Species Composition of Some Beech-Maple Forests

Fagus grandifolia	62.1	48.0	72.4	40.1	10.2	62.1
Acer saccharum	31.0	23.0	7.4	21.5	73.2	31.2

Source: Braun, E.L., *Deciduous Forests of Eastern North America*, Hefner Publishing Company, New York, 1972.

abundances of the two dominant canopy species in six randomly selected examples of beech-maple forests studied by Braun. The data show American beech as typically more abundant than sugar maple, the differences being partially attributable to soil type, climate, and age of the forest. The remaining differences in abundance are probably stochastic (i.e., other factors) in nature. On average, however, American beech is more abundant than sugar maple in such forests and it may be asked if such data would undermine the stochastic species hypothesis.

Given periods of time that are suitably long, especially with the possibility of significant climate change, the answer could be "not quite." Bearing in mind the behavior of random walks as described earlier in this section, it may well be the case that the dominance persists long enough to imply the absence of a complete stochastic orbit for some species in such late-successional communities. On the other hand, there is no question that, as far as the 125 biosurveys in the meta-study are concerned, the forest communities that appear in the meta-study followed the same J-shaped pattern as other communities did. To explain this seeming anomaly, we ask if the notion of a random walk can accommodate (if not "explain") such long-term patterns. Or, as Den Boer (1990) puts it, "Random walks of densities do not exclude the possibility that local populations can persist for some centuries."

A classic experiment on lengthy random walks by the statistician Feller (1968, Vol. 1) revealed unexpected anomalies that still surprise people today, even those with a statistics background. When people encounter the phrase "random walk," they are apt to imagine a sequence that goes up and down in an irregular fashion, but without thinking about long-term behavior. Feller's random walks were constructed like those of today. A random sequence of +1s and −1s contributed to an ongoing sum, just as the ups and downs of a species in the MSL system contribute to its population size. It therefore does no violence to Feller's experiments to speak of his time series as the ongoing record of a population (with negative values allowed of course). The statistic of interest in this experiment was the number of times the population crossed zero, going from positive to negative values or conversely. During the experiment in question, Feller observed some eight changes of sign with one run of 7804 steps. He remarks that, "Sampling of expert opinion revealed that even trained statisticians expect much more than…eight changes of sign." Very long runs are therefore not unexpected and, by a simple extension of the theory, may involve quite large populations being sustained for surprisingly lengthy periods. The underlying statistical law governing this phenomenon is called the arcsine law (Morters and Peres 2010). If one interprets such a result as a species undergoing annual periodic reproduction, for example, a 7000-year run of high abundance might thus be captured by what is a purely statistical phenomenon, albeit produced by a confluence of environmental mechanisms.

At this point, we come to an important intersection in the road to understanding how hyperbolic theory works. Classically, for example, dominance in beech-maple forests has a simple biological explanation. Returning for a moment to this example, the reason that beech and maple so often predominate in this type of forest is that the shade tolerance of younger individuals of these species enables them to survive under a closed canopy, whereas other species tend not to survive this stage unless there is an opening overhead. Not all changes in population are explained by such regular phenomena, although many are. Other, less regular phenomena tend to operate locally, sporadically, and as the ultimate result of factors and events that have nothing to do with the species in question. Without much exaggeration, it could be said that hyperbolic theory "explains" nothing but accounts for everything. In such a context, the ultimate linking study would no longer be deep, but broad. Instead of observing the effect of single factors on a species of interest, one could (theoretically at least) study all possible factors affecting birth and death.

On the other hand, factors leading to dominance may themselves be "temporary" over varying stretches of time due to a variety of influences, which may be viewed as "effectively random," as described in Section 1.4. Think glaciations, continental drift, large meteor impacts, orogeny, coronal mass ejections, epicontinental sea incursions, extended volcanism, and so on.

7.2.5   CYCLIC CHANGES IN ABUNDANCE

Slow seasonal changes in weather or other gradual factors that affect abundance can be viewed as an ocean swell that lifts a pattern of foam up and then down again, the pattern evolving independently of the swell. The J-curve for a community of insects in a temperate biome might well stretch out horizontally as the warm weather intensifies, with insect populations increasing almost daily in abundance. With the onset of cold weather, the J-curve shrinks back to smaller dimensions. Of course, how one counts the insects under these circumstances may alter the picture considerably. Insects may deal with cold weather in a variety of ways. In North America, among the Lepidoptera alone, for example, the Monarch butterfly migrates south to Mexico, the Mourning Cloak overwinters in the adult form, the Black Swallowtail overwinters as a pupa (chrysalis), the Wooly Bear overwinters as a larva, and the Eastern Tent Caterpillar overwinters as an egg (Layberry et al. 1998).

If one counts only the adults in a given population, the count during the winter months would drop to zero, except for those species that overwinter as adults or migrate away. If one counts different life stages, however, nonzero counts will be maintained. Indeed, if one counts the egg form, the population would appear to surge during late summer and fall, when most of the eggs are deposited. Egg mortality would gradually reduce the count until spring, as with other forms. In cases like this, one does not see the typical ups and downs of the stochastic vibration that has been invoked (by implication) for so many other biota. However, a simple alteration of the underlying stochastic system will reproduce the result without changing the hyperbolic outcome.

Section 10.4 describes a variant of the weakly stochastic system in which the probabilities of birth and death, instead of fluctuating in a neighborhood of equality, follow what might be called a seasonal probability curve. Thus, in the early spring, let us say, the probability of birth, previously less than the probability of death, begins to climb rapidly as eggs begin to hatch, even as the probability of death begins an abrupt decline. By the end of the warm season, the two probabilities are approximately equal, following which the probability of birth begins a slow decline. Obviously, one may impose any regimen one likes on seasonal changes, as long as the two probabilities come into balance at least once a year.

At the same time, one must take account of various cycles in other populations such as the predator–prey cycles as described by the Lotka-Volterra equations and seemingly apparent in trapping data (Leigh 1968) and from other sources. This seems the appropriate place for the following remark on the famous Hudson's Bay population records for the Lynx and Varying Hare. These have been used by a great many authors as an example of a predator–prey cycle. Had those authors taken a closer look, however, they would have discovered that no such cycle is involved. The original literature on this subject makes it clear that the Lynx are not driving the cycle, but rather cleaning up the corpses of hares already dead from epidemics that sweep through hare populations whenever they reach a critical density (MacLulich 1937). However, the result is much the same.

Predator–prey cycles have little effect on samples of communities since typically only two species are involved and the cycles themselves are somewhat irregular. On the other hand, one may implement the MSL system with two species A and B and reinterpret births and deaths as follows. An increase in A at the expense of B may be interpreted as predation. However, an increase in B at the expense of A may be interpreted as the failure of A to find B, dying as a result, with a subsequent increase in B's population. This system does not produce cycles of any regular kind. Indeed, the classic experiments of Gause (1934) with (real) protist predators (*Didinium*) and prey (*Paramecium*) also failed to produce any cycles.

8 The Meta-Study
A Review

The empirical foundation for adopting the J distribution as a universal descriptor of abundances in natural communities rests on the meta-study introduced in Chapter 1 and explained more fully in this chapter. The study used 125 randomly selected biosurveys (or field surveys) as random samples of their respective communities (as defined by their respective authors). Each sample histogram was fitted to a J distribution having the same mean abundance and initial bar height as the sample. The resulting 125 chi-square scores were then compiled into a new histogram that portrayed the collective shape of the scores. If that histogram of scores was a relatively close match to the chi-square distribution itself and (more importantly) if the mean chi-square score matched the score expected under the null hypothesis, it would not only imply the presence of the J distribution in the data collectively but also exclude from that role any proposal that scored significantly higher.

Theoretical support for the procedure just described uses the concept of a perturbation of a theoretical distribution in its discrete (histogram) form, essentially a sample (with replacement) of intensity 1.0 of a theoretical histogram. The hypothesis under test is that the 125 sample datasets (as embodied in their respective histograms) may be regarded collectively as perturbations of the J distributions to which they were fitted. In more detail, one starts with a sample distribution F' from one of the 125 field studies. A theoretical distribution F with the same mean abundance and initial value F'(1) is readily determined by the method described in Section 8.3.2. However, if the field histogram has n abundance categories, the number of degrees of freedom for the fitted distribution will be $n - 2$, since the J distribution has two parameters. A chi-square goodness-of-fit test at $n - 2$ degrees of freedom is then performed on the pair (F, F'). The score of the test is then added to the list of scores thus emerging, as earlier. This approach to the data is justified by the body of theory that begins in the next section. A final step involves harmonizing the scores by converting them to equivalent scores at 10 degrees of freedom. In the final steps, the grand average score is computed and the scores compiled collectively into a large histogram that can be compared directly with the chi-squared distribution itself.

8.1 BACKGROUND: THE CHI-SQUARE THEOREM AND TEST

As stated in standard treatments of distribution theory (Kotz 1970) and (Neuts 1973), the theorem on which the chi-square goodness-of-fit test is based comes in two parts. The first part treats a random variable X having a normal density N(0, 1).

Theorem A: If x is a random variable having a normal density function with mean 0 and standard deviation 1, then the random variable x^2 has a chi-square distribution (X^2) with one degree of freedom.

The second theorem extends the first to a more general form:

Theorem B: If X_1, X_2, X_3, ..., X_n are independent random variables having the normal density function $N(0, 1)$, then the sum

$$X_1^2 + X_2^2 + X_3^2 + ... + X_n^2$$

has a chi-square distribution with n degrees of freedom.   ■

The chi-square test is based on the second theorem. The test statistic consists of an algebraic measure of the difference between an empirical distribution F (derived from a field sample, say) and a corresponding theoretical distribution F′ with which it will be compared. Readers may assume that the empirical distribution consists of a species-abundance histogram while F represents a J distribution with suitable parameters, as described in Section 4.4 and in Section 8.3.2.

$$\text{score} = \sum (F(k) - F'(k))^2 / F(k) \tag{8.1}$$

The summation typically runs from $k = 1$ to m, the latter number resulting from a grouping operation that is part of the test: an abundance category in F′ with fewer than 5.0 species is grouped with subsequent categories until the total number of species thus subsumed equals or exceeds that number. Thus, in a particular instance, k might run from 1 to 254, but after $k = 6$, the values of F′(k) fall below 5.0. Subsequent categories must then be grouped into new, larger categories with a larger expected sum. For example, a new category might then consist of 7 and 8 together, yielding a sum F′(7) + F′(8) that equaled or exceeded 5.0. This sum would then be compared with the sum F(7) + F(8) to provide the next contribution to the score formula. Continuing the process, the next set of combined categories might extend from 9 through 15 before the sum of F′-values equaled or exceeded 5.0. This time, the two sums to be compared would embrace seven categories each. The last category might consist of the sum of all the remaining F′ and F values (the vast majority of which would be zeros). The null hypothesis for this test would be that the empirical data follow the J distribution.

The square root of the general term of Equation 8.1 may be written as

$$X_k = (F(k) - F'(k)) / \sqrt{F(k)}.$$

Under the null hypothesis, the random variable X_k would have mean zero (owing to the subtraction) and a standard deviation of 1, owing to the division by the square root of F(k), the theoretical frequency.

Under the null hypothesis and by Theorem B, the test statistic would have the chi-square distribution as pictured in Figure 8.3 and 8.5. The actual value of the statistic in a particular test is then compared with a critical value for the chi-square distribution at the appropriate number of degrees of freedom. Tables of such values are readily available.

If the test statistic exceeds the corresponding critical value, it will be found in the tail of the chi-square distribution with an associated probability density that represents the chance that one would err in declaring the null hypothesis invalid. As normally applied, therefore, the test is actually negative, being more reliable in rejecting hypotheses than in accepting them. If one formed the same test statistic for another distribution with a shape more or less like the J distribution, one might well find oneself having the same degree of "confidence" in respect of it, as well. Under the null hypothesis, if one continued to perform the chi-square test on other empirical distributions from the same source, the resulting scores would form a distribution that would converge on the chi-square distribution itself. This observation lies at the foundation of the method about to be explained.

8.1.1 AN ILLUSTRATION OF THE CENTRAL METHODOLOGY

The method can be illustrated by a simple example involving the uniform random distribution. One can generate samples of this distribution in a manner that exactly mimics the conditions of chi-square theory. In other words, generation proceeds by sampling the parent distribution at intensity 1.0 (with replacement). This guarantees a normal distribution in each column for the resulting number of species and so the chi-square theorem applies. The result of the process will be called a *perturbation*.

Since Pearson's theorem applies to any distribution whatever, we can illustrate it with the uniform random distribution U[5] × 50, as shown in Figure 8.1.

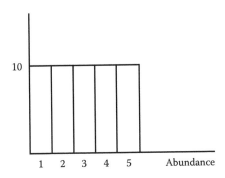

FIGURE 8.1 A uniform source distribution.

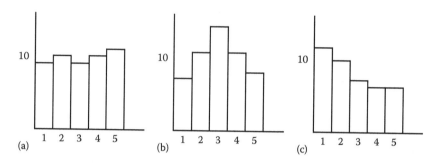

FIGURE 8.2 Perturbations of the uniform distribution may produce different shapes, e.g., rectangular (a), triangular (b), or sloping (c).

The distribution shown in Figure 8.1 has five categories and a total of 50 items (interpreted as species, if one wishes) distributed uniformly among them. Using U as the source distribution, we may generate as many perturbations as we like by the sampling procedure described earlier.

Figure 8.2 shows three examples of perturbations thus generated. These were selected from among the first dozen examples that were generated by computer, seeming pathologies being relatively common. They illustrate the point that visual inspection of samples can be misleading. While Figure 8.2a seems to be a more typical sample, odd-looking samples abound. For example, Figure 8.2b and c appear to have originated in unimodal and sloping distributions, respectively, yet both happened to be perturbations of the uniform distribution in this instance. Were one testing a tent-shaped distribution in this manner, the histogram of Figure 8.2b might well be a perturbation of that distribution as well. Perturbations are in no sense unique to a source but represent that source collectively in a more precise manner.

If we now generate 100 perturbations of the theoretical curve U[5] × 50 and compute a chi-square score for each one in comparison with U[5] × 50, we may compile the scores into a new histogram in which the height of each bar is the number of scores that fall into that category. Such an experiment was performed using a computer program that also computed the chi-square score for each perturbation. The experiment was run for 100 trials. The resulting histogram appears in Figure 8.3 with the corresponding chi-square distribution superimposed on it.

As expected, the fit is fairly good for a compilation of chi-square scores, the more important result being the average score of 4.958, close to the expected 5.000 for an average fit under the null hypothesis. The theoretical chi-square values (the curve) may be compared with the experimental values, rendered as bars. Naturally, there is some variation in the latter, owing to the usual statistical fluctuations. In fact, a direct comparison with the meta-study result (Figure 8.5) shows approximately the same degree of raggedness.

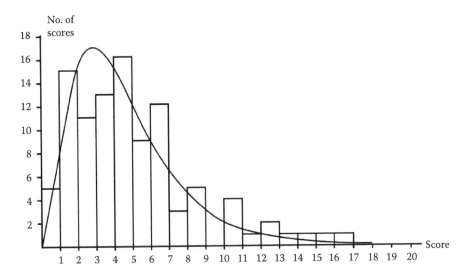

No. of scores

FIGURE 8.3 Histogram of chi-square scores for uniform distribution.

8.2 APPLYING CHI-SQUARE THEORY TO MULTIPLE-SOURCE HISTOGRAMS

In Section 8.1, it was shown how chi-square theory works in the standard case of multiple perturbations of a single source. In this section, we apply it to single perturbations of multiple sources, so to speak. The space of all possible perturbations and sources is represented schematically in Figure 8.4. Here, the horizontal plane represents the parameter space of the J distribution, with each grid point representing

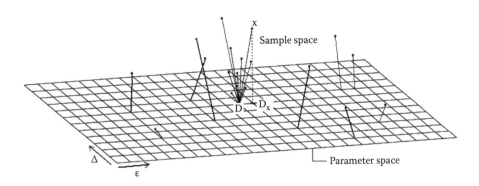

FIGURE 8.4 Parameter space and sample space.

a pair of parameter values. Values of the parameter epsilon are represented by the cross-page axis, while the parameter delta follows the other horizontal axis "into" the page. If all such distributions had, say, 10 degrees of freedom, every possible J distribution with 10 degrees of freedom would be represented by a point in this two-dimensional parameter space.

In the vertical dimension, above the plane, we may locate all 125 sample histograms as follows. Given any field histogram X, we may determine the best (chi-square) fit Dx and use it to assign parameter coordinates to X. The vertical coordinate will simply be the chi-square score of the fit, with poor fits being located higher above the coordinate plane than better ones. In this manner, each of the 125 sample histograms may be located in the sample space above the plane. At the same time, we may carry out the $F_a - \mu$ procedure (Sections 5.1 and A.2.8) on X to determine a J distribution (called the *source*) of which it may be regarded as a perturbation, as explained in Section 8.1. Since the source distribution so determined is normally different from the best fit distribution, the line joining X to its source distribution is usually slanted. In this representational scheme, the fitting procedure plays no other role than to enable us to assign a position to the perturbation in sample space; the figure as a whole is intended only as a visualization tool for the equivalence between two treatments.

In this illustration of the equivalence, each of the lines in the cluster is identical to one of the isolated lines, having the same length and orientation. This would of course not happen if one attempted the actual experiment. In fact, it would be highly unlikely that any of the lines so matched. However, there would be nothing to prevent such a coincidence and therein lies the point of the figure; given a clustered set, there would be an equivalent isolated set and conversely. The two sets of lines, one clustered, the other dispersed, illustrate the application of chi-square theory to the many tests comprising the meta-study. Carrying out many tests on different field histograms is equivalent to perturbing a single J distribution many times and testing the results. Since the scores associated with the clustered source follow the chi-square distribution, so do the scores associated with the isolated points. The only question that remains is how close the average score will be to the number of degrees of freedom of the source, namely, 10 in the case at hand: that would be the expected optimum score under the null hypothesis.

8.2.1   THE SCORE CONVERSION PROCESS

The whole point of worrying about degrees of freedom is to ensure that the score conversion process is consistent with the underlying chi-square theory that gives meaning and validity to the meta-study. The actual scores in the meta-study vary considerably, as would be expected in any case. The next step is to convert a score taken at d degrees of freedom to the equivalent score at 10 degrees of freedom.

Consider then the chi-square process involving c abundance categories (as defined by the chi-square process itself). The J distribution has two parameters, and their presence reduces the number of degrees of freedom by two, yielding $c - 2$ degrees of freedom. A third degree of freedom is lost by virtue of both source and distribution perturbation having exactly R species so distributed. Thus we have, finally, that

$$d = c - 3$$

in all 125 cases.

The foregoing arguments lean heavily on the assumption that the results of a chi-square test with some other number of degrees of freedom can be made equivalent to the results of a test on 10 degrees of freedom. I will illustrate the equivalence operation with a simple example in which the chi-square table plays a key role.

In Table 8.1, I have excerpted a small portion of two rows from a standard chi-square table of critical values. In the process, I have reduced the decimal fractions from five decimal digits to three for the sake of simplicity. As explained in Section 4.4.1, the q values represent the probability that one is correct in rejecting a chi-square score that is higher than the corresponding critical value. Suppose, for example, that one has conducted a chi-square test with 6 degrees of freedom on a particular field histogram, obtaining a score of 4.721 in the process. Note that this score exceeds the critical value of 3.455 yet falls short of 5.348.

Because it exceeds the lower critical value, one could say that the score is too large to be "accepted" as having the J distribution at the level of $p = 0.750$. In other words, one could reject the null hypothesis (that the histogram has the J distribution) with a probability of 0.750 of being right. The corresponding probability p of wrongly rejecting the null hypothesis is 0.250, since the complementary probability has the relationship $p = 1 - q$ with q. If the score exceeded 5.348, on the other hand, the probability of rejecting it incorrectly falls to 0.500. As I pointed out in Section 4.2, one must have a rather higher score in order to reject it at the highest level, where the probability of being wrong has fallen to $q = 0.050$. At this level, the corresponding p value is 0.950, and one says that one has 95% confidence that a rejection was correct.

The foregoing explanation serves only to refresh the reader's memory about the chi-square table and how it is normally used. Here, I shall use the table to provide an index of equivalence, so to speak. What score at 10 degrees of freedom corresponds to the score of 4.721 at 6 degrees of freedom?

The simplest approach is mildly approximate. Linear interpolation treats the probability function between the two critical values as a straight line, whereas the

---

TABLE 8.1

Critical Values for 6 and 10 Degrees of Freedom

Q-value	0.750	0.500
6 df	3.455	5.348
10 df	6.737	9.342

function is slightly curved over the domain spanned by the two values. The discovery of an equivalent score x at 10 degrees by this method involves a simple proportion. The estimated score x occupies the same position within the interval [6.737, 9.342] as 4.721 does in the interval [3.455, 5.348]. The calculation results in the value x = 8.479.

The chi-square score at 10 degrees of freedom corresponding to a score of 4.721 at 6 degrees of freedom is, under this scheme, 8.479. The two scores may be said to be equivalent under linear interpolation. The continuous versions of the curve formed by critical values as a function of p value is very slightly curved upward over the domain exceeding p = 0.500 and very slightly downward below that value. The interpolated values are therefore slightly too high in the first case and slightly too low in the second case. Such errors would have an impact on the final average by introducing consistent inaccuracies over the domains in question. Under the null hypothesis, this would drive scores up slightly.

8.3   THE META-STUDY

The following sections describe stages in the overall procedure of collecting and analyzing the data (field histograms) that comprise the meta-study. The data collection stage involved a random selection of field studies from the extant literature at the time the meta-study was carried out. The meta-study compiled scores not only for the J distribution in relation to each field sample but also for the log-series distribution. The next section describes the fitting procedure in each case.

The third section explains the process used to convert scores to their 10-degree of freedom equivalents, compiling them into two histograms, one for the J distribution and one for the log-series distribution. Of all the extant proposals for the theoretical distribution that best matches the J distribution, the log-series is clearly the closest in shape. If the J distribution is the correct overall descriptor of abundances in natural communities, it should have a better (i.e., lower) score than any of the others, including its closest relative, the log-series.

8.3.1   THE DATA COLLECTION PROCEDURE

Data for the meta-study originated in some 125 papers, as published in a wide variety of journals (see "Meta-study References" in the Bibliography). Each paper was selected at random from the University of Western Ontario library by graduate students who had been given a method for selecting journal volumes at random and, within those volumes, for selecting a random paper that contained a biosurvey or list of species found in a certain community, along with their abundances. The only nonrandom element was an occasional intervention on my part to include a specific group, hitherto missing from the study by chance. It can be said, however, that selection of papers within such groups followed the same randomizing procedures. In all cases, however, papers that listed fewer than 30 species were excluded from the study, as were papers that addressed only "common" species within a group. The papers were collected under these criteria prior to any analysis of their contents. None were rejected beyond the stage of collection and the preliminary assessment just described.

From each paper, a species-abundance table (field histogram) was extracted. If there was just one set of data, those data were used. If there was more than one set of data, several cases arose. In one kind of case, the biodiversity of a disturbed area was compared with that of an undisturbed area. In this case, the undisturbed area was chosen. If the biosurvey was broadly based, with samples from more than one station, the first sample containing 30 or more species was chosen. Some biosurveys sampled more than one kind of community (e.g., wasps and spiders). In this case, the larger community was chosen for analysis.

8.3.2 Testing the Data against Two Distributions

In all datasets thus selected, the abundances were compared, via the chi-square test, with the predictions made by both the J distribution and the log-series distribution. The latter distribution is the theoretical proposal closest to the J distribution in overall shape, so it made a useful basis for comparison. If it did not do as well as the J distribution, the other theoretical proposals would fare worse.

The parameters for either distribution were selected as follows:

For the J distribution, the $F_a - \mu$ method of curve-fitting was used. In this method, the mean abundance μ of the field histogram, along with the number of species F_a in the minimum abundance category were used to estimate the corresponding values of ε and Δ in order to obtain a candidate "source" distribution. (See the transfer equations in Appendix A.2.8.) While some of the resulting fits were better than average (less than the number of degrees of freedom), other fits were worse. Such a result was not only consistent with chi-square theory but also demanded by it in this context. The 125 scores were then compiled into a "meta-table," as shown in Appendix B.

For the log-series distribution (see Appendix A.2.7), there is only one way to select a candidate source distribution with the same mean, the log-series being a one-parameter distribution. The parameter value c and constant α, not being independent, were both derived from μ, the mean abundance. Thus, when it came to comparisons, both distributions were on an equal footing, both being fitted to curves based on the mean abundance and adjusted with their respective degrees of freedom. Again, some of the scores were suboptimal, while others were higher, as was the case with the J distribution. Indeed, it was fully expected that some of the log-series fits would be better than their hyperbolic counterparts, and this happened from time to time. Again, 125 chi-square scores were recorded, this time for the log-series.

8.3.3 Converting and Compiling the Scores

All chi-square scores for both sets of tests were then converted into the equivalent scores at 10 degrees of freedom. (This corresponds to a common p value, for those familiar with the chi-square test.) The purpose of this normalization process was to make the results of the 250 tests comparable within the data for each distribution, as well as between distributions. A paired interval test was employed to detect any significant difference between the two sets of scores.

Since any fit of a J distribution to empirical data will result in a specific value for the parameter Δ, that value automatically becomes a prediction for a canonical position of the maximum abundance. The comparisons were made in the form of a percentage ratio $\Delta'/\Delta \times 100$, as described in Section 5.1. We note in passing that detecting any difference between the two sets of test scores by eye alone would be completely impossible. Averages, standard deviations, and other basic statistics were indispensable to the study.

8.4   EXPERIMENTAL RESULTS OF THE META-STUDY

The results of the meta-study support the claim that the collective score for the J distribution was optimal, or as near to optimal as one could reasonably expect, given the variability in the scores themselves. As described in the earlier discussion of this result in Section 1.5, the score of 10.0 is the expected score for a family of perturbations under the null hypothesis at 10 degrees of freedom. The average chi-square (normalized) score of the meta-study was 10.43, which is reasonably close, given the high variance (SD = 4.64) of the scores. The degree of closeness to the chi-square distribution itself was calculated in another way.

We could of course avoid all the foregoing complication and employ a much simpler (but cruder) method of computing an average score: if one computes the percentage ratio of each individual chi-square score to the corresponding number of degrees of freedom, then computes the average, the result in the present case will be 99.8, slightly below the expected 100.0. Further confirmation arises from how well the theoretical values of Δ matched the empirical ones. The average ratio, expressed as a percentage, is 101.6. Rarely does one get closer results, even for a correct theory.

Figure 8.5 displays a histogram of chi-square scores superimposed on the chi-square distribution itself. It hardly follows the expected distribution exactly, especially as it is shifted somewhat to the right, having a higher mean. The variability

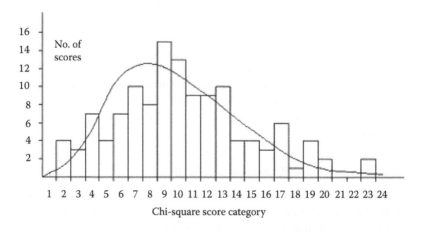

FIGURE 8.5   Distribution of chi-square scores superimposed on chi-square distribution.

in bar height may be compared with the variability in Figure 8.3, which we already know to be the kind of result to expect under these conditions.

The overall test score for the log-series distribution was 13.56 (SD = 9.40), well above the score for the J distribution, but within a standard confidence interval of it. To tease out systematic score differences for the two theoretical distributions, a paired interval test was applied. In this test, the statistic is formed as the sum of individual squared differences of pairs of scores, one pair for each set of biosurvey data. The compiled sum of these differences resulted in the interval [0.64, 4.89]. In other words, the mean difference between the paired scores lay within this interval with a probability of 99.9%. The statistical interpretation can also be phrased as follows: "We would err with a probability of 0.001 in claiming that the paired differences are bounded away from zero." In other words, there is a residuum of difference that cannot be accounted for as random. The better score of the J distribution in this context is apparently the result of it being the better descriptor of species-abundance field histograms, in general.

Although the (slightly) higher-than-expected average chi-square score of 10.43 and the (slightly) higher-than-expected Δ-prediction of 101.6 are about as close as can be expected, one is still free to speculate that the two results are related in the following sense: there was a slight tendency for those histograms that fit the source distribution less well also to have maximum abundances that had ratios of Δ'/Δ higher than 100.0. A reverse finding would be slightly more surprising. The small errors introduced by the score interpolation procedure, as described at the end of Section 8.2.1, may have had an influence on the average chi-square score, as well.

The tests carried out as part of the meta-study were very time-consuming, involving hundreds of hours to complete. There seemed little point in including tests of the lognormal and other distributions for the following reasons. First, of all the extant proposals for theoretical species abundance distributions, the log-series is closest in general shape to the J distribution, both being essentially hyperbolic functions. Although the log-series is a reasonable approximation to the J distribution for some purposes, the insertion of a convergent series (as suggested by R. A. Fisher to Corbet and Williams in 1943) amounts to what physicists would call a "fudge factor," meant to force the area under the log-series curve to have a finite value. Inaccuracies introduced by this factor, although slight, appear to show up in the significantly different overall test scores for the two distributions.

As for the lognormal distribution, the situation is basically hopeless. When used with the correct "veil curve" (Dewdney 1998b) and the standard abundance axes, the lognormal distribution would yield average normalized scores one or two orders of magnitude higher than the results of the meta-study, putting the lognormal out of the running. One reason for the poor fits would be the fact that the theoretical lognormal curve would tend to have a low value in the lowest abundance category, while the histogram with which it is compared would have a rather high one. Moreover, at slightly higher abundances, the situation is reversed: where the typical field histogram has declined to fewer species, the lognormal is peaking. As I have pointed out in Section 4.2, the use of logarithmic axes in such a comparison would prove nothing; when applied to the J distribution, the result is a similar unimodal distribution, as shown in Figure 2.6.

9 Fossil J-Curves

Although it might seem an odd thing to do, a biologist conducting a biosurvey could count not only the individuals per species collected, he or she could also count the number of species per genus, erecting a histogram for such counts. In this case, the first column would indicate the number of genera that had just one species in the area under survey, the second column would indicate the number of genera with two species, and so on. Indeed, for the sake of examining the patterns thus produced, it would not be necessary for the biologist to confine the count to his or her sample, or even to the area under survey. The same exercise could be carried out over a much wider area with the help of a field guide or specialized monograph. Not stopping at species per genus, the biologist could also count the number of genera per family, indeed the number of any lower taxonomic level per a fixed higher level.

In all cases, the same shape would emerge, sometimes smooth, sometimes less so. The experience would be eerily similar to plotting the original biosurvey of species over individuals. Why on Earth should a J-curve show up again in a taxonomic setting? I will try to answer that question toward the end of this chapter. The first question to answer, however, is whether it does actually appear. Do the taxonomic abundance data mentioned in the previous paragraph match the J distribution to the same degree as the biosurvey data studied in Chapter 8? This time, the Kolmogorov-Smirnov (K-S) test will be applied to the matching process. The comparison results in strong support for the hypothesized answer to the question.

A mathematical inquiry into the presence of the J distribution in taxonomic data explores two quite different evolutionary scenarios starting in Section 9.3. It has proved difficult to make a convincing argument under the uniformitarian point of view (Ridley 1996, p. 51), but much easier in the punctuated equilibrium scenario (Gould 2009).

9.1 BACKGROUND OF THE PROBLEM

Up to this point, the main focus of this monograph has been present-day communities, sets of species, with each species being a set of individuals. But a consideration of communities in the distant past leads us to (a) a taxonomic structure that goes beyond these two categories and (b) a concept of "community" that expands to continental and even global scales. For the purposes of this chapter, the definition of a *taxonomic community*, whether past or present, will mean the set of all individuals living at a certain time and sharing a single common ancestor, without being very specific about the time. If that ancestor were identified as the founding member of a new family, then the community in question might consist of all individuals living today, for example, and descended from that ancestor.

If we take only the traditional taxonomic levels (omitting suborders or superfamilies, for example), we find a hierarchy of sets, each set being subdivided by its predecessor.

individual < species < genus < family < order < class < phylum < kingdom

We have already seen that when the species in a community are sorted into abundance categories, a J-shaped curve emerges. But when the members of a given taxon are sorted into subtaxa, the same shape seems to emerge. For a random example of the phenomenon, we may distribute all the known families of North American herpetofauna (Conant and Collins 1991) into those having just one genus, then those with two, then three, and so on, obtaining Figure 9.1 in the process. Indeed, the phenomenon extends not only to consecutive hierarchies but also to any pair at all: If we choose two different taxonomic levels from the hierarchy and examine the distribution of the higher taxa into their abundances in terms of the lower ones, the same shape of histogram emerges.

One of the first observations of the J-shape in taxonomic distributions was made by our early hero, C. B. Williams (Fisher et al. 1943), the British biologist. Summarizing his research at Rothampstead over the 1940s and 1950s, Williams (1964) writes as follows:

> One general principle, or mathematical pattern, has several times been discussed. In almost every classification that has been proposed the number of genera with a single species is greater than the number with two, the number with two greater than with three, and so on. If one plots such a classification in the form of a frequency distribution we get a hollow curve, not unlike a hyperbola, which immediately recalls the similar pattern already discussed for the relative abundance of species.

Over the last century, others have noticed the "hollow" shape of taxonomic distributions, notably Willis and Yule (1922), Burlando (1990, 1993), and Chu and Adami (1999). In most of the cited research, a power law is proposed, compared with a few

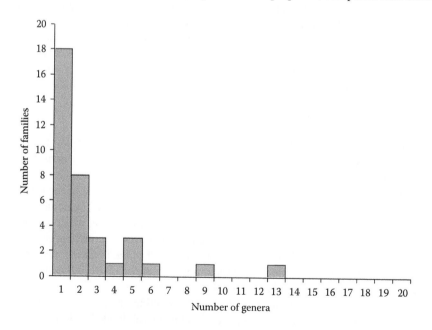

FIGURE 9.1 Numbers of families of herpetofauna containing a given number of genera (one family with 35 genera occurs off-page).

datasets, and justified in terms of branching processes in (abstract) taxonomic trees. The simplest power law is from the first authors cited. One may apply it to species and genera, as follows

$$G(k) \, \alpha \, k^{-g}. \qquad (9.1)$$

Here, α represents a proportionality relation, k is a taxonomic abundance in species, G(k) is the number of genera having k species, and g is a constant lying in the interval [1.4, 1.6]. We note in passing that a power law with exponent near unity is very close to a J distribution, the underlying hyperbola having the formula k^{-1}. The exponent 1 is just outside the stated interval. Perhaps, the difference is due to the nonconvergence of a power law with a unit exponent.

Chu and Adami (1999) propose a somewhat more general scheme called the Galton-Watson (Shi 2010) process that produces a sequence of generations (whether of species, genera, or other taxa) governed by a simple probability distribution p that for each taxon and each possible number k of "offspring" (a lower taxon) gives the probability p(k) of that many offspring at the next "generation," presumably a rather long stretch of time. The authors compare the results of the process under slightly different values of the average number of offspring, from a pure power law to a negative exponential distribution. The results are compared with one dataset drawn from the paleontological record using a K-S test (to their credit) and the fits are about as good as one might expect from a distribution that resembles the J distribution. In any event, the authors are under the same impression as those referred to in Section 1.4, that a single set of data, whether from the present or the past, makes a reasonable test for a hypothesis.

Once it is suspected that the phenomenon of J-shaped taxonomic distributions is rooted deeply in the past, it might also be suspected that a uniform stochastic dynamic produces such distributions. Over extended time and space, genera themselves may have undergone a stochastic vibration of sorts, losing species and gaining them by extinction and speciation, respectively, with the same dynamic applying to higher taxa, as well. However, this proposition has been difficult to establish directly, as explained in Section 9.3.1, and may not even be true. Perhaps, something like a Galton-Watson process may be applied in such a circumstance. At the same time, the idea of punctuated equilibrium (Eldridge and Gould 1972) opens the way to an entirely different explanation.

9.2 PRESENCE OF THE J DISTRIBUTION IN TAXONOMIC DATA

The statistical tests applied to the taxonomic data used here not only evaluated the J distribution as a descriptor of taxonomic abundances but also were used to evaluate scale-restricted and less formal sources of taxonomic data for bias in the resulting statistical scores. Table 9.1 lists the source of information for each taxonomic group used in the study, as well as the extent of geographic and taxonomic coverage. In the table, I have indicated taxa by one-letter codes (e.g., "s" = species, "g" = genus, etc.). This makes it possible to calculate the number of tests available from the data.

TABLE 9.1

Sources of Taxonomic Information

Taxonomic Group and Range	No.	Geographic Coverage	Reference
Plantae [f to c]	3	Global	ITIS 2002b
Gymnospermatophyta [s to f]	3	North America[a]	Kartesz 1994
Pteridophyta [s to f]	3	North America[a]	Kartesz 1994
Animalia [o to p]	3	Global	ITIS 2002c
Mammalia [s to o]	6	North America	Hall 1981
Reptilia & amphibia [s to o]	6	North America	Conant and Collins 1991
Aves [s to o]	6	Global	Sibley and Moore 1990
Pisces [g to o]	3	Global	Nelson 1984
Testudines [s to f]	3	Global	Iverson 1992
Insecta [f to o]	1	North America	Borrer and White 1970
Arachnida [s to f]	3	E. North America	Caston 1972
Collembola [s to f]	3	Global	Bellinger et al. 1996–2003
Pogonophora [g to f]	1	Global	Ivanov 1963
Ciliophora [g to o]	3	Global	Lynn 2007
Foraminifera 1988 [g to f]	1	Global	Loeblich and Tappan 1988
Fungi [f to c]	3	Global	ITIS 2002a
Bacteria [s to p]	3	Global	Holt et al. 1994
Life [c to p]	1	Global	Wilson et al. 1973

Note: Total datasets: 55.

[a] Including Greenland.

For example, [s to g] means "species to genus," and only one set of data, species within genera, is available. On the other hand, [s to f] or "species to families" makes three datasets available: species within genera, species within families, and genera within families. I have placed the number of datasets so derived immediately after this notation.

For some groups it was not possible to obtain the coverage desired. For example, in Class Pisces (Nelson et al. 2016), the ostensible coverage was [s to o]. But for many genera, this global synopsis gave no precise figure for the number of species therein, providing only estimates: "about 60." Another example, the Fungi (ITIS 2002a), illustrates the use of online data in this inquiry. The Integrated Taxonomic Information System (ITIS) was still under construction at the time of this particular research. It yielded only [f to c] coverage, owing to having many genera absent from its lists. At higher levels, with no apparent omissions, the data were assumed to be complete. The remaining ITIS-based tests involved plants and animals. In the latter case, all data pertaining to phylum Ciliophora were omitted, as these do not properly belong to Kingdom Animalia. More recently, an authoritative taxonomic table relating to phylum Ciliophora was found (Lynn 2007).

The bacterial data, drawn mainly from *Bergey's Manual* (Holt et al. 1994), listed species within genera and the latter within "Groups." I interpreted this word as synonymous with phyla or divisions, as in standard treatments on bacteria, the names being largely the same. (See Margulis and Schwartz 1982 for a summary.) At the same time, the manual gives no taxonomic abundance of species for 21 genera, stating that the determination of the species within these genera involves tests too advanced for the manual.

The last taxonomic group, namely, "life" itself, involved an attempt to compile all the classes within the approximately 92 phyla of life on the Earth. In addition to the source listed in Table 9.1, I used the general reference (Margulis and Schwartz 1982) and several general web-based resources such as the Tree of Life Web Project (TOL 2001) and the UK Systematics Forum (UKSF 2004). Phyla not covered in these sources included Sipunculan worms (Cutler 1994) and several protistan phyla (Margulis et al. 1990).

9.2.1 THE TEST METHOD

In the meta-study of species abundance data, I used the chi-square test, but for this investigation, I switched to the K-S test, partly because of its alleged sensitivity, but also to demonstrate the confirmatory approach with a completely different goodness-of-fit test. I could have used a chi-square test, of course. When a multiple-data test of the sort described here produces pass–fail ratios that are as close to the ones predicted by the K-S test under the null hypothesis (see Tables 9.4 and 9.5), there is little room for any other distribution to play a role. To put it another way: any other distribution that succeeds as well with this taxonomic abundance data would have to be very similar to the J distribution.

Data from all the sources in Table 9.1 were put into histogram form for each taxonomic abundance pair. In the case of species within families, for example, I counted the number of species in each family for the group in question, creating a standard histogram (as in Figure 9.1). The data so organized were subjected to the K-S test for goodness of fit as follows.

As explained in Section 4.4.2, the test statistic uses maximum differences:

$$D = \max\left\{\left|F'(k) - G'(k)\right|; \; k = 1, 2, \ldots n\right\}.$$

In the context of this chapter, $F'(k)$ will be the number of higher taxa that contain k instances of a lower taxon, as in the number of families, for example, that contain k species. The function $G'(k)$ will be the corresponding value of the J distribution function. How close is the predicted number of higher taxa to the actual number? Is the maximum difference great enough to cause D to exceed the critical value of the test? In the context of a comparison between a taxonomic histogram and the J distribution, the critical value amounts to a kind of envelope around the J distribution. If the histogram values all fall inside this envelope, the comparison is given a pass.

In the K-S test, there is a critical value for each size of sample and for each level of significance. Respective scores (D values) achieved by each taxonomic abundance histogram were compared with the critical values at the 95%, 90%, and 80% levels of significance. A histogram that passed at the 95% level of significance might well fail at 90% or 80% since in those cases, the test is more lenient in regard to rejection. If one wants only a smaller probability of being wrong in rejecting a particular histogram, the appropriate critical value is higher; if the difference D exceeds even this value, the histogram is even more likely not to fit. It follows then, that one would expect in many cases to find acceptance ("pass") at one level of significance while finding rejection ("fail") at a lower one.

The respective scores for each taxonomic abundance histogram were compared with critical values in order to decide whether the data passed or failed the test. If the histograms collectively follow the underlying K-S distribution, one would "expect" the number passing at the 95% level to be 95% of the histograms thus tested. Similarly, 90% should pass at the 90% significance level and 80% at the 80% level. D values were also converted into equivalent D values at a nominal—but fixed—sample size of 25. Each D value was normalized by applying a correction factor of $\sqrt{N/5}$, where N is the sample size of the dataset. Having normalized scores made it possible to compare scores evaluated at different sample sizes.

Two questions arise naturally in this context:

1. Does the common origin of taxonomic histograms within a source group influence their independence?
2. Does it make any difference that field manuals, rather than formal synopses, were sometimes used?

To answer these questions, the normalized test scores were subjected to simple tests that compared average scores and variances of the data in question with the overall results.

9.2.2 RESULTS OF THE STUDY

Over all 55 taxonomic datasets, the (normalized) average K-S score was 0.166 and the average (normalized) variance was 0.0064. The question of statistical independence within source groups was addressed by examining whether the scores within groups tended to be clustered. The contributions to overall variance from within groups ranged from small values to numbers in excess of the overall variance of 0.0064. One may check this important observation by examining the results of the study, as listed in Appendix C. Variances within the data presented in the three largest tables are displayed in Table 9.2. In the first two cases, the variance exceeds the overall variance and in the third case it is slightly less.

The numbers imply that the common origin of data within a group did not reduce its overall variation. It may therefore be assumed that the datasets are statistically independent.

Results for field guides versus synopses were not so clear cut. Although for some of the groups covered, such as birds, field guides would be taxonomically complete at a continental scale, others, such as Insecta (Borrer and White 1970), could not be, owing to the vast number of insect species in comparison with the number of pages in a typical book (or website). As it happened, field guides came in with an average

TABLE 9.2
Variances within the Three Largest
K-S Score Tables

Table	Variance
First	0.014
Second	0.023
Fifth	0.006

score of 0.117, somewhat below the overall average score of 0.166. Although it could be said that using field guides favored the theory under test, the small number (11) of datasets derived from field guides may have resulted in coincidentally low numbers. A brief test of this possibility involved the score for a field guide not used in the study (Whittaker 1996). North American mammal data from this source yielded a test score of 0.159, a bit shy of 0.166. The latter guide, however, is substantially complete in taxa. Table 9.3 displays the K-S scores for all data sources, both field guides and synopses, representing continental scale groups.

The results of the full null hypothesis test appear in Appendix C, where the individual scores are organized on a taxon-within-taxon basis. Thus, "genera within

TABLE 9.3
Contributions of Continental-Scale
Groups to K-S Scores

Taxonomic Group	K-S Score
Pteridophytes	0.194
Angiosperms	0.239
Arachnids	0.272
Herpetofauna	0.110
Mammals	0.287
Pteridophytes	0.270
Angiosperms	0.281
Arachnids	0.123
Herpetofauna	0.102
Mammals	0.130
Mammals	0.122
Pteridophytes	0.322
Angiosperms	0.128
Arachnids	0.032
Herpetofauna	0.108
Mammals	0.157
Mammals	0.099
Insects	0.072

TABLE 9.4

Comparison of K-S Test Score Percentages with Expected Percentages

Number predicted	95%	90%	80%
Number passed	92.7%	87.3%	81.8%

TABLE 9.5

New Percentages with One Test Difference in Each Category

Number predicted	95.0%	90.0%	80.0%
Number passed	94.5%	89.1%	80.0%

families" means that for each taxonomic group with both genera and families represented, the corresponding test score is included under the given heading. Raw test scores that did not pass the K-S test are in italics in the table.

In two of the tests, both involving extremely large numbers of taxa, the fits were exceptionally poor. In probing the reason for this, I discovered that the high score had less to do with a failure to match the shape of the proposed distribution per se and more to do with rather extreme variability from one abundance category to the next. By reducing all abundance numbers to 10% of their original values, the effective sample size was reduced to lie within the upper end of the range of the other sample sizes. New lower scores emerged, effectively damping the variability. However, such data modification imposes a certain caution on conclusions influenced by either of the tests in question. Table 9.4 summarizes the outcome of the main experiment.

The agreement between the predicted and actual percentages is fairly good, with (percentage) differences of −2.3%, −2.7%, and +1.8%. Within each significance level, some 55 tests were conducted.

To make the point of how close to optimal the foregoing results are, Table 9.5 shows what the results would be like if just one more test had passed in each of the first two score categories, along with one less passing in the third category. Even for a correct theory, one will hardly get closer than the percentages listed in Table 9.5 when 55 such tests are carried out.

9.3 EVOLUTIONARY ORIGIN OF THE J DISTRIBUTION

As discussed in Section 7.1, a "community" can be any size, from local to global in extent. In global- or continental-scale communities, the corresponding J distributions would, in many cases, inflate to enormous sizes, with values of Δ running into numbers that are several orders of magnitude greater than are normally found in samples. In many cases, such maximal communities would have high capacities, as defined in Section 2.1.1, ensuring relatively large numbers of species with relatively small populations. Such a context is assumed in this section.

In the next two subsections, two quite different possibilities for the origin of the J distribution in taxonomic data will be explored. In both scenarios, we will assume a uniform taxonomic judgment to be in play to assess speciation events, as though such a taxonomist had gone back in time to decide the kind and degree of genomic change.

9.3.1 Stochastic Genera

In providing sound theoretical reasons for the presence of the J distribution in taxonomic data, we first adopt a uniformitarian (Ridley 1996, p. 51) point of view and ask whether a new version of stochastic behavior is possible, this time focused on genera, rather than species. The "stochastic genus hypothesis" would assert that, over time, genera would be as likely to gain a species as to lose one. However, the argument in this case is complicated by several possibilities for the kind of speciation involved, not to mention a problem arising from timetables for extinction. The main modes of speciation may be summarized as follows (Raup and Stanley 1978).

Case 1: In successional speciation, a species has been undergoing phyletic change (genetic drift) and has just crossed a taxonomic boundary that results in the "appearance" of a new species, accompanied by the "loss" of its rather similar ancestor. The net change in the population of the genus is in this case zero.

Case 2: In the splitting of a lineage, a species s undergoes mutation (by whatever means) and remains on the scene along with its offspring species. The net change in this case is different, according as either of the following subcases apply:

Case 2a: The new species is not different enough, taxonomically speaking, to be counted as a new genus and the net change in G's population of species increases by unity as a result, since the new species still belongs to G.

Case 2b: The new species is different enough to be counted as a new genus and the net change in G's population is a decrease by unity.

To complete this simplified list of possibilities, one must include a final case:

Case 3: The species becomes extinct and the population of G decreases by 1.

However, the final clincher for the argument supporting the J distribution applied to species was that every individual that is "born" must sooner or later die, almost always within a limited period of time (Section 7.2.1). In the very long run the presence of such a stochastic dynamic might well be operative at the level of genera but complicated by the absence of a fixed, a priori limit on how long a species might be expected to survive; species have no life expectancy, so to speak (Van Valen 1973). Any argument attempted along these lines, whether applied to genera, families, or any higher taxa, cannot be ruled out as a candidate explanation waiting in the wings for the arrival of the appropriate insight.

9.3.2 EPISODIC SPECIATION

The second explanation involves rather short periods of time in which much of the speciation for a given epoch takes place. This mode of evolution works much better, providing an explanation of how new families may appear, along with genera that are J-distributed, among them. The argument can then be extended up the taxonomic tree.

Consider a genus G, along with all of its species, at some point during the Earth's history. Each genus is embodied not only by its species but also all of the populations of those species, the latter having the J distribution among the species. Figure 9.2 illustrates the situation in terms of a set diagram.

The outer circle of the set diagram represents the genus G, while the inner circles represent the various species composing the genus, while each dot within a species circle represents an individual organism belonging to that species. The widely differing populations of dots of course reflect the J distribution.

Suppose now that a sudden episode of speciation takes place that affects all individuals in a manner that is essentially uniform random. Some new individuals are hardly different from their parent(s). Others, produced by a Type 2a process, are different enough to be counted as new species. The remainder, produced by a Type 2b process, become not only new species but also new genera, as well. These appear in the figure as black dots instead of open ones. Statistically speaking, the number of individuals in each of these three categories (previous species, new species, and new genera) continues to reflect the J distribution in respect of the original species—circles.

The original genus G, since it now contains new genera as well as new species, becomes promoted, so to speak, to family status. However, since the original genus G may still contain species that are not substantially different from those previously present, it remains as a genus, now within the new family. Meanwhile, the family containing the original genus G that now contains a new family either splits into two

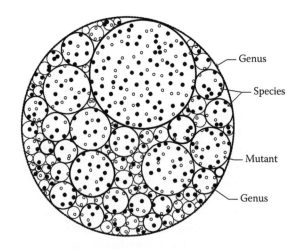

FIGURE 9.2   Speciation events in genus G for an episodic scenario.

families or becomes a new order in its own right, and so on. In general, the number of speciation events within a circle will be proportional to the size of the population inhabiting it, and in this manner, the relative population sizes are inherited, in a manner of speaking, by the new genera.

The foregoing description proves nothing, being both oversimplified and somewhat speculative, as well. But it does show potential for a rather neat explanation for how the J distribution might be inherited.

Did such episodes ever actually happen? The best example we have of such episodic evolution is provided by the concept of punctuated equilibrium (Eldredge and Gould 1972) and later by Gould (2009). As many biological readers are aware, one of the main empirical supports for the theory of punctuated equilibrium comes from the many complete fossiliferous stratigraphic sequences found around the world that (a) span millions of years and (b) typically show very little phyletic change in the organisms represented there. Even in the present day, according to Raup and Stanley (1978), there should not be as many "living fossils" as we have, were phyletic evolution as common as episodic speciations: These groups derive their name from the fact that they have persisted for long spans of geologic time with very little morphologic change. The important point is that taxa that have persisted for long intervals at low species diversities have almost invariably evolved very little.

As a possible source of the kind of sudden change that might trigger a great many speciations in a short time, one might invoke a variety of cosmic events, not to mention volcanoes and earthquakes. Solar events such as solar flares and the more energetic coronal mass ejections might be powerful enough to break through the shielding effect of the Van Allen Belts, flooding the Earth with ultraviolet (UV) radiation, x-rays, or gamma rays for days or weeks. During pole reversals (averaging millions of years apart), the Van Allen Belts break down for months or even years, during which life on Earth has little protection from the same kinds of radiation.

Over the last decade, a new scenario has been steadily gaining ground as the most serious explanation of punctuated equilibrium itself. According to solar physicist Henrik Svensmark (2012), the solar system has experienced several explosions of "nearby" (in the local galactic arm) supernovae in past epochs. He demonstrates convincingly that each galactic event has coincided with a spike in the rate of speciation. Among the effects produced by such explosions are (short-term) a flood of high-energy ionizing radiation at all frequencies of the electromagnetic spectrum, as well as neutrons and cosmic rays (Genet et al. 1988). These are so massive and powerful that the Van Allen belt offers no protection at all. Even the sun's powerful magnetic field, which normally shields all four of the inner planets from cosmic radiation, is of little avail. When a cosmic ray strikes the Earth's atmosphere, it induces a shower of high-energy x-rays. Any organisms not immediately killed by the flux would undoubtedly experience chromosomal damage resulting in a high incidence of death, infertility, and abnormal offspring, many of them infertile and/or destined for extirpation.

At the same time, there would be enough nonlethal mutational effects to provide a rich source of genetic novelty to engender a wholesale bloom of speciation. In one stroke, an extensive dieback is accompanied by a flood of new species, many of which might become extinct in time, yet many surviving to add new taxonomic layers, so to speak. The findings by Svensmark were to some degree anticipated by

Tsakas and David (1986), who, writing on the subject of neighboring supernovae, also postulated a powerful dose of ionizing radiation. One of the new factors introduced by Svensmark is a cooling effect of cloud formation engendered by cosmic rays in particular (Svensmark 2012). The resulting ice age would presumably be responsible for most of the extinctions. As mentioned previously, the sun's magnetic field plays an important role in shielding the inner planets (including the Earth) from cosmic radiation. This field follows the well-known solar cycle in which the solar magnetic field reverses itself every 11 years. However, there are times when the field simply fails to develop to its normal strength, remaining weak enough to open the magnetic door, so to speak, to powerful fluxes of cosmic radiation and particles. The absence of sunspots during the "Little Ice Age" (Rafferty 2016), which started around 1315, are interpreted as an indication of a weak solar magnetic field, one that resulted in extensive, long-term cloud formation and much cooler global temperatures. Of course, one cannot rule out this companion factor as an important cause of much deeper and longer ice ages known to geologists.

According to a handbook on radiation biology published by the International Atomic Energy Agency (2010), there can be little doubt about mass extinctions in such scenarios. As for speciation events, Rothschild (1999) describes the effects of UV light as follows:

> One of the best-known consequences of radiation exposure is DNA damage. DNA damage may be accurately repaired, result in a mutagen, or cause death. Without genetic novelty, there would be no variation available for evolution to occur. Mutation is the source of genetic novelty. Thus an extremely important role of UV radiation in evolution is that of mutagen.

Rothschild then goes on to make the important distinction between "DNA damage" and "mutation," describing some of the mechanisms in each case. Laboratory experiments on the effects of radiation on mutation of course abound, often carried out on the vinegar fly, *Drosophila melanogaster*. From the late 1920s up to the present (Muller 1927, Wallace 1958, Stephenson and Metcalfe 2013). A wide-ranging and extensive summary of biological effects of a wide variety of radiation sources will be found in *Radiation Biophysics* (Alpen 1998), where the effects of radiation on plant, animal, and microbial organisms are summarized. The long history of such research has made it clear that all such forms may have a role to play in evolution. That role plays out equally well in both the phyletic and episodic scenarios.

There is little available information on the effects of radiation on wild populations, except for isolated incidents such as the Fukushima nuclear accident at Fukushima in 2011. In that case, a local butterfly population showed mutational changes attributed to high radiation in the area (Hiyama et al. 2012).

9.4 EXTINCTION AND SPECIATION IN NATURAL AND ARTIFICIAL COMMUNITIES

Extinction rates in experimental stochastic communities such as the multispecies logistic (MSL) system appear to be comparable with actual rates, although examples of the latter are not easy to find. As proxy for such real data, we must turn to experimental populations As for speciation, there is a good reason to suspect that the multitude of small, somewhat isolated populations of a given species (all part of the same metapopulation) operate as a structural guarantee of that possibility, as will be seen.

9.4.1 Extinction Rates in the MSL System

A great deal of research has been devoted to the problem of understanding extinction rates in natural communities, mainly through work with experimental populations. It is well understood that population size and "demographic stochasticity" are the dominant factors in determining the fates of individual populations (Griffen and Drake 2008).

Experiments with the MSL system have revealed that, depending on the value of ε or, equivalently, the minimum abundance, an isolated community will inevitably begin to lose species, owing to stochasticity within smaller populations. Yet the value of N will remain more or less constant. This can only mean, in this context, that the value of ε will continue to climb, while the minimum abundance (in the canonical sense) creeps slowly higher. As might be expected, extinctions become increasingly rare over time, as the community settles into a quasi-equilibrium.

The actual extinction rates in the MSL system appear to be consistent with what we observe in real communities. Perhaps the most famous example of the latter was described by MacArthur and Wilson (1967) in their well-known study of island biogeography. When the island of Krakatoa exploded in 1883, all (visible) life on the island was presumably extirpated by lava and hot ash. But within a few decades, some 30 species of birds had established themselves on the island, along with many plant species, all of them presumably immigrants. At each visit, ecologists found that a few species were no longer present but that new ones had arrived, more or less preserving the equilibrium number. MacArthur and Wilson estimated an average annual extinction rate of about 1.15% of species per year.

Accordingly, I set up the MSL system with 30 species of "birds" having an average population of 200 individuals each (an estimate), then ran it to equilibrium (6000 cycles), at which point, of course, some populations were greater than 200 and others considerably smaller. At this point, the extinction switch was automatically turned on, and over the next 1000 cycles, the artificial community typically lost nine species, a function of this particular combination of richness and abundance.

To discover roughly how much time in the life of a bird community 1000 cycles might represent, I used the following approximate reasoning: One cycle involves 100 iterations of the basic operation of trophism and reproduction (birth and death). If 100 individuals reproduce in that time, it would represent all the females in a total population of 200 birds (breeding pairs). We may therefore estimate that each cycle was equivalent to turnover in a single species and that 30 cycles could therefore be taken as equivalent to one breeding cycle for all the birds on the island. Taking 30 cycles as equivalent to a single year, 1000 cycles would represent 1000/30 = 33.3 years. Thus, in the pseudo-avian equivalent of 33.3 years, the MSL system produced nine extinctions, with the heaviest losses initially and declining losses thereafter. The initial loss rate was about 0.9% per year. If new species had been supplied to the system at this rate, the total number of species would of course remain approximately the same, and the behavior of the system under these circumstances could hardly be distinguished from its behavior with the extinction switch simply turned off.

The average turnover of 1.13% on Krakatoa noted by MacArthur and Wilson is certainly in the same order of magnitude as the 0.9% produced by the MSL system under comparable circumstances. Indeed, given the relatively unsophisticated nature of the basic MSL system, not to mention the back-of-the-envelope approach taken here, the agreement is acceptably close. For example, since Krakatoa is well inside the tropical zone, one could allow two breeding cycles per year and arrive at a turn-over rate of 1.8%, somewhat higher than the estimate of 1.13.

9.4.2 SPECIATION IN STOCHASTIC COMMUNITIES

The fields of ecology and evolution are often both represented in biology departments around the world. The reason for this is simple. Evolutionary theory must draw heavily on ecology in order to reconstruct the biota of past epochs. Limits to ecological knowledge automatically becomes limits to evolutionary knowledge.

The problem of speciation is a core topic within paleontology and evolutionary biology. The study by Rice and Hostert (1993) summarizes research up to the date of that publication, where several theoretical models of speciation are evaluated in the light of breeding experiments, with organisms ranging from vinegar flies to fish. It may be that, in spite of varying degrees of support offered by experimental evidence, all proposed mechanisms have a role to play, even if some play only a minor role. This situation offers an ironic contrast to the plethora of proposals for species abundance distributions, where none may have a role to play. In any event, most of the proposed mechanisms depend on isolation of populations and acknowledge, implicitly or explicitly, that smaller populations are at some point necessary.

A large population can often be viewed as a metapopulation, with subpopulations distributed over a large area, perhaps even continental in scale. The hyperbolic theory tells us that many of these subpopulations will be rather small but that, collectively, they may be large, qualifying them in a probabilistic sense for a speciation event. Of large populations that do not consist of isolated pockets, Mayr (1970) writes:

> The real problem of speciation is not how differences are produced but rather what enables populations to escape from the cohesion of the gene complex and establish their independent identity. No one will comprehend how formidable this problem is who does not understand the power of the cohesive forces in a coadapted gene pool.

When a small local population receives a mutation, it has an opportunity to spread to all members of the population. Most often, the mutation is deleterious or makes no difference to the survival of the small population. Even if the mutation is advantageous, the population may well become extirpated. But if the population survives and ultimately grows, it carries the mutation with it. Eventually, the larger population replaces adjacent populations or remixes with them, introducing the new gene(s) into the larger pool and adding to the richness of the collective genome, not to mention enhancing its survival.

Most of the populations in a J distribution are relatively small, no matter how large the community is. In fact, an inspection of the population mean reveals that the great majority of species have smaller-than-average abundances, with many of those smaller still. Although not all of these populations are isolated in any reproductive sense, some of them may well be. Such circumstances favor not only the phyletic version of speciation but also episodic versions. It must be added, in light of the conjecture in Section 9.3.2, that in the event of a catastrophic "punctuation," most populations, both large and small, would be decimated, leaving scattered small populations in a great variety of situations. In such cases, there would be little to fear from a "cohesive gene pool." In any case, the low abundance end of the J distribution may be seen not only as the doorway to extinction, but also as the cradle of evolution.

10 Summary of Theory and Open Problems

This chapter includes a summary of the main theoretical and empirical results that support the hyperbolic theory. It includes, as well, a list of open problems and lines of research that, in the author's opinion, show promise for future developments. The "big picture" that emerges from this view of natural populations has a satisfying quality of closure in view of the role played by a conic section at the center of things: while planets follow elliptical orbits, species appear to follow hyperbolic ones. The planetary orbit is deterministic, while the species orbit is stochastic; a population may advance and retreat in unpredictable fashion, sometimes less, sometimes more. But according to the stochastic species hypothesis, it will trace out a hyperbolic shape all on its own, given enough time, as a result of all the biotic and physical interactions with the world of which it forms a part. Owing to its simplicity, variants of the hypothesis are inevitable, as in Section 10.4.

The hyperbolic shape of the species abundance distribution is inevitable, given the stochastic species hypothesis. Although the racetrack analogy described at the end of Chapter 1 may be helpful for some readers, others may find it too fanciful, demanding something a little more realistic. Consider then a sealed cylindrical tube containing a gas at equilibrium, at uniform temperature and pressure throughout. If one end of the tube is heated while the other end is cooled, a temperature gradient develops along the tube, with molecules at the hot end moving very quickly, while those at the cool end have slowed down. Thanks to the equilibrium process I am about to describe, the pressure continues to be uniform throughout the tube. It follows that the gas at the cold end is denser than the gas at the hot end. Equilibrium may be explained in terms of an imaginary partition placed anywhere along the tube. For purely statistical reasons, the number of molecules traversing the partition in one direction must equal the number crossing in the other direction; any time an imbalance in the flow occurs, the side with more molecules must eventually give more of these up to the other side of the partition. That is what equilibrium means. Now if we think of the molecules as species, their position in the tube is proportional to their temperature, just as the position of a species along the abundance axis is proportional to the rate of stochastic vibration, so to speak. The density of the gas is inversely proportional to the temperature, as it happens, and an inverse-linear function is nothing more than a hyperbola. The number of molecules per unit volume at temperature (or position) t is proportional to $1/t$.

The problems that have presented themselves as this book was being written have outstripped the author's ability to provide solutions. Thus, a section devoted to open problems has been added to outline the leading edge of the larger inquiry. Some of the problems may have simple solutions and some may turn out ultimately to be intractable. They are, for the most part, amenable to the approach used so often in the foregoing pages. A mathematical formula or expression is (a) proved as a theorem and/or (b) either established or illustrated by appropriate computer experiments and (c) sometimes supported by rather large randomly selected sets of data. By the same token, the problems that have proved amenable to solution have appeared quickly enough to make publishing them all as part of one unified monograph more efficient than publishing them as separate articles.

On the theoretical side, the mathematics used cannot be described as especially difficult. Students with the normal preparation of two years of calculus, along with a course in applied statistics, should have no trouble following the mathematics, although they may find the details occasionally tedious (as I have).

Some readers may have noticed that the word "model" rarely appears in the foregoing pages. The word has been used in so many contexts, both within biology and without, that it drags a certain nebulosity along with it and too often involves explanations that are little more than guesses. I avoid the word as mildly inappropriate in the context of "exact methods." Although almost self-explanatory, this term requires some explanation. The cries to make population biology (not to mention all of biology) into an "exact science" are frequent enough to require no documentation. A special meaning must be given to the term when applied to an essentially statistical subject like population biology. In that context it simply means statistical exactitude coupled with clear hypotheses and proofs of statements. Statistical exactitude of course means the consistent use of interval statistics or error bounds to accompany all derived measurements or estimates. The hyperbolic theory expounded here is but one example of an exact method.

10.1   SUMMARY OF RESEARCH

This monograph proposes a new theoretical distribution as the underlying pattern for the abundances of species in communities *and* in samples of them. The distribution has been derived theoretically through an analysis that is (a) primarily mathematical and (b) based on a very simple statistical hypothesis, that of the stochastic community. All conclusions flow inexorably from these two sources. The resulting pattern has the shape of a pure hyperbola, the J distribution.

The formula so derived has been tested against 125 randomly selected field samples. The meta-study described in Chapter 8 (see also Dewdney 2003) is new to the ecological literature in two ways. First, the number of tests applied to a single theoretical proposal far exceeds the number ever carried out previously. Second, the tests so applied, namely curve fitting, have been more stringent than any previously used. The results of these tests give the strongest possible support, supplemented by two further confirmatory tests of how well the theory fits the data, namely the average ratio of delta values coming in at 101.6% and the average ratio of chi-square scores to degrees of freedom achieving 99.8%. Nothing less is required to run the blockade created by the error of misplaced generality (Section 1.4). Because the methodology employed by theorists up to this point in time has become a more or less entrenched practice, an approach as novel as the one described here may well be rejected out of hand by reason of its unfamiliarity. It may take several years for the approach to be widely recognized as valid and the results conclusive.

This assessment may sound overblown, but only in its historical context. An approach to sampling based on goodness-of-fit tests should have (and could have) been taken more than 50 years ago. The long delay in the adoption of such methods can be attributed to misconceptions arising from the error of misplaced generality. When a biologist has a field sample to compare with a theoretical distribution, the sample data are often cast in a species-abundance histogram or chart of abundances. Under such circumstances, it has been the standard practice in other sciences to prepare a histogram of corresponding theoretical predictions and to carry out a standard statistical goodness-of-fit test. A professional library research technician was unable to locate a single article in this particular area of population biology that employed such a test (Galsworthy 2004). This was a surprising result, even after my own failure to find such an article among the 200-odd papers I reviewed in the course of the meta-study.

Such tests would have at least provided a metric or measure of similarity that would have made precise what were purely visual assessments of similarity that paved the way for vague judgments and uncertain conclusions. For example, researchers could have rejected some distributions and "accepted" others by the criteria of the chi-square test just for starters. Such methodology would not only have enhanced analysis, it might also have focused more attention on the role played by sample variability in how well field data matched the theoretical distributions under test. It might have made it clear that the superiority of one theoretical distribution over another for a handful of samples from the field is almost meaningless and that a great many would be needed for the purpose.

Given the inadequacy of the traditional evaluation process, the simplest starting assumption is that all communities follow the same template. The picture that emerges from the research reported here implies a constant turnover in the degree to which various communities will fit any fixed distribution. The turnover is perfectly natural, moreover, entirely predictable as a phenomenon and (almost) entirely unpredictable in detail. As the tests reported in Chapter 8 indicate, the collective fit of the 125 communities to the J distribution is so close that no other proposed distribution can play such a role without being identical to it.

To review these results, the stochastic species hypothesis holds for simple stochastic systems almost by definition and holds for more general versions of such systems, as demonstrated in Section 7.1.2. The appearance of a binomial shape in field data (population changes), as demonstrated in Section 7.2, is fully expected as the natural outcome of the stochastic species hypothesis. The J distribution itself is mathematically implied by the stochastic species hypothesis and shows up with the expected score in the extensive chi-square tests of Chapter 8.

In contrast to "unified" theories of biodiversity, the theory presented here is "uniform" in the sense that it seems to apply everywhere, providing estimation methods for the species richness of communities, sample accumulation formulas, species–area relationships, overlap and similarity for both samples and communities, not to mention a formula for rank abundance curves. The theory also applies established mathematical theorems such as the Pielou transform and the Weierstrass Uniform Approximation Theorem to prove the General Sampling Theorem, as well as Pearson's chi-square theorem to establish the meta-study protocol used to prove the presence of the J distribution in biosurvey data.

For reasons yet to be fully understood, the J distribution (read J-curve) also shows up in taxonomic abundance data, as described in Chapter 9. The one explanation given leans heavily on the concept of punctuated equilibrium, although I have not given up looking at the possibility of a "stochastic genus" hypothesis. Such an approach, however, would lean equally heavily on a gradualistic (Darwinian) view, which, these days, seems rather less likely as the principal driver of evolutionary change. Whatever the mechanism of its emergence, the evidence for presence of the J distribution in taxonomic data is strong.

10.1.1 EXAMPLES AND COUNTEREXAMPLES

The research reported here also includes many worked examples to clarify procedures such as richness estimation methods, overlap calculations, species–area curves, and so on. It also includes counterexamples to demonstrate the fallibility of various concepts, including biodiversity (as defined by various authors), data representation schemes, and some proposals for species distributions. Counterexamples were also used to demonstrate the need for estimating sampling intensity as defined in Section 3.3.2.

10.2 A GUIDE TO FIELD METHODS AND THEORY DEVELOPMENT

A correct description of species abundances in natural communities immediately becomes the core insight that ushers in a wide variety of field methods and modes of analysis—a toolkit for those who develop theory, as well as those who collect field data with a view to describing the communities that the samples represent.

It has already been made clear that one needs to know both the underlying distribution and the intensity of one's sample in order to take full advantage of the methods described in Chapters 3 through 7. The distribution that underlies field data can be reconstructed by the methods developed there. In Chapter 3, for example, it is shown how to estimate the intensity of one's sample. The field biologist may then use the method of Chapter 4 to arrive at estimates for the parameters ε and δ that lurk in his or her data. The methods are exact in a statistical sense, converging (over multiple samples) to correct relative average values. It is also important, when a biologist takes several samples, whether of the same community over time or different communities over space, to know how similar the samples are, based on the overlap methods developed in Chapter 4. Finally, the most important application of all is to arrive at estimates for the richness of the community under investigation. Once again, those estimates have been shown by computer experiments (Chapter 5) to give statistically accurate results.

In the course of showing how the hyperbolic theory might be used, I have developed or adapted additional mathematical tools for those who work in mathematical biology. The canonical sequence developed in Chapter 2, for example, provides a window on abundances in natural communities with the "Noise" factored out, so to speak. The deeper one looks into a large community via sampling intensity, the more spaced out the sampled abundances become, even as the minimal abundance slowly edges away from seeming extirpation. A more general tool is the integral

I have called the Pielou transform in Chapter 3. It tells us precisely how the process of random sampling affects the shape of the underlying distribution in passing from community to sample. Using the transform, I was able to provide a mathematical demonstration of the essential fallacy in the "veil line" concept. The "line," as such, is neither vertical nor even straight, but a slanted sigmoidal curve that gives an entirely different result when the nonshowing species are subtracted. In any case, when the J distribution is subjected to a logarithmic transformation, it also produces a truncated, unimodal, bell-shaped curve, as demonstrated in Section 2.2.4.

Perhaps, the statistical tool of greatest significance employed in this monograph is the application of Pearson's theorem to a multitude of chi-square tests, instead of just one. In this context, acceptance and rejection of the null hypothesis become somewhat irrelevant individually. All that matters is the distribution of scores in comparison with the chi-square distribution itself. The validity of the method is beyond doubt, and its proper interpretation gives students of the method a chance to evaluate communities of data, so to speak. Auxiliary tools assist not only in the chi-square analysis but also in the estimation methods of Chapter 5: the transfer equations play a role in both methods, as well as the computer programs, SampSim and StoComm, as described in Chapters 5 and 7 respectively. The interface will be found at the author's website at http://www.csd.uwo.ca/~akd/ inside the page "Research in Population Biology." Appendix A contains a summary of these tools, including the address of a website where they may be found in a graphic user interface.

A different goodness-of-fit test was used to probe for the presence of the J distribution in taxonomic data. Some 55 datasets, derived from taxonomic tables embracing a span of biota nearly half as large as the meta-study itself, turned out to follow the J distribution in the same collective sense. The Kolmogorov-Smirnov test provided a simpler criterion in which percentages of passes or failures became the collective score for taxonomic abundance histograms. It is possible that the J distribution is, in some manner, inherited from abundances of ancient times. The simplest explanation leans heavily on punctuated equilibrium, possibly the result of cosmic disturbances such as supernovae in our own galaxy (Svensmark 2012). However, the possibility of an explanation that fits more comfortably with neo-Darwinian theory is still "on."

10.3   OPEN PROBLEMS AND PROSPECTS

Not to treat the present theory as a closed book, it is important to highlight areas where further work is needed and to encourage students of the subject to pursue the lines of research indicated in this chapter. The amount of progress already made varies from one problem to another.

How does the parameter ε increase with intensity r?

In Section 6.1 it is made obvious that the parameter ε increases in value as the sampling intensity goes up, but it is not yet obvious that it increases in linear fashion. It would be useful to the development of the hyperbolic theory if this could be confirmed, whether experimentally or theoretically. Of course, the trend may not be linear, but something more complicated.

Is there an explicit formula for the species accumulation curve?

In Section 6.2, two formulas for species accumulation appear, one for accumulation with replacement and one (Hurlburt's formula) for accumulation without replacement. Neither formula has an obvious closed form, but one may exist, nevertheless. Even an approximate formula (equipped with suitable bounds) might be useful.

Can the J distribution arise under phyletic evolution?

It would seem, as remarked in Section 9.3.1, that the most hopeful route to such a result would follow a path parallel to the stochastic species hypothesis. Is it possible to prove by invoking a "stochastic genus hypothesis," so to speak? Can one make the argument without taking into account the unbounded longevity of species? Perhaps the branching argument of Chu and Adami (1999) might be adapted in a hyperbolic context to give a useful result.

What is the expected time-series profile for population variability?

In Section 7.2.2, it was shown that one sign of the presence of stochastic behavior in the population of a single species is the appearance (over an extended period of regular sampling) of a binomial or normal distribution of changes in abundance from one sample to the next. Such normal-shaped shapes emerge readily in bird count data (Robbins et al. 1986). The appropriate way to demonstrate the presence in such data is to average out the histograms over 100 count differences, 200 differences, and even 300. A chi-square fit of the normal distributions to the resulting histograms will show steadily decreasing scores if the stochastic species hypothesis is correct. I can find no other explanation in the literature for the emergence of a normal distribution in this context.

Can multiple environmental factors be compiled as probabilities?

It has been claimed in these pages that the apparent stochasticity in abundances results from myriad interactions between a species and its environment (Sections 1.6 and 7.2.3). Up to this point, however, the "myriad" influences have yet to be described in any detail. Let us examine, just as an example, the birth and death of a beetle, without being specific about the species. Beetle eggs deposited in the soil may or may not even hatch, depending on moisture, temperature, the absence of egg predation, and other factors. Any probabilities attached to such events would have to be determined experimentally, although some of the data may already be available (e.g., Jacobs et al. 2014). They are apt to be small, however. Backing up, one could also study the probabilities connected with egg production, success of the mother finding a suitable laying site, survival of the mother up to the time of laying, and more. A more or less complete list of factors affecting the birth of a beetle into larval form might well run into the hundreds, with many possibilities being so remote as to be excluded by most biologists as irrelevant. But the cumulative effect of numerous tiny probabilities would startle most people who are unfamiliar with the phenomenon.

For example, suppose I have a list of factors that will each harm a developing egg with probability 0.001. For the egg to hatch (a reasonable point to describe as a "birth"), *none* of the improbable events that might kill the egg must occur. In other words, we must look at the product of a great many *complementary* probabilities that are close to 1.000. If we multiply 25 probabilities 0.999 together, a probability of 0.9753 that none of the factors will affect the developing egg will result. Multiply 50 of them and the probability drops to 0.9511. In other words, the probability that none of the factors just referred to will harm a typical beetle egg is 0.9511, whereas the probability that at least one of them will harm the egg is the complementary probability of nearly 0.05. Thus in a clutch of 100 eggs, one might expect some 5 of them to be affected by one or more of the factors in question. This simple example is merely intended to illustrate the important role played by a myriad of separately "irrelevant" factors.

10.4   STOCHASTIC SYSTEM AS RESEARCH TOOLS

As described in Section 1.2, the multispecies logistical (MSL) system began as a dog-eat-dog interaction among mutually predatory species. Once it became apparent that the system had another, altogether different interpretation (without altering the underlying dynamic), the way to further generalizations lay open, as described in Section 7.1.1. The following versions of the basic system warrant the term "detail hungry" as applied to the basic system. At the very core of all the stochastic systems explored here lies the inner dynamic that guarantees statistically the emergence of the characteristic J shape. It results, remarkably enough, from the tendency of abundant species to change their abundances more quickly than less abundant ones. In any case, the following elaborations of the basic dynamical system would qualify as valid research vehicles, whether in a context of specific "models" or for the purpose of exploring general behavior over a wide range of initial conditions.

10.4.1   THE STOCHASTIC SYSTEM

The stochastic system generalizes the one-for-one replacement scheme of the MSL system by a probabilistic rule that allows births and deaths with equal probabilities, as described in Sections 1.2 and 7.1.2. The basic dynamical cycle selects an individual (not a species) at random and either removes it from the system (death) *or* it adds a new individual to the species to which the selected individual belongs (birth). The probabilities of birth and death are fixed and equal.

10.4.2   THE WEAKLY STOCHASTIC SYSTEM

As described in Section 7.1.2, the weakly stochastic logistical system generalizes the stochastic one by allowing the probabilities of birth and death to wander from equality, as long as a long-term average equality is maintained. In this system, one observes with interest that probabilities themselves may be given probabilities. In both systems, the total number of individuals is no longer fixed but varies more or less normally about a characteristic value.

10.4.3   THE SEASONAL STOCHASTIC SYSTEM

Is it possible to custom-fit the J distribution to communities that undergo seasonal changes? Section 7.2.1 describes seasonal variations within a community of temperate zone insect populations, for example. A time-varying probability curve that extends over an annual cycle may have any shape one likes, but with the death probability higher over the cold season and the birth probability higher over the warm seasons. The only *sine qua non* would be that both probability curves should have equal integrals annually. The cyclic stochastic logistical system exists at present only in the proposal form. In adding detail to the system, one is free to specify a different probability curve for each species in a community, it being necessary only to ensure that the curve densities sum to unity. All three systems produce the hyperbolic curve when histograms are taken post-equilibrium.

10.4.4   THE COMPARTMENTALIZED TROPHIC SYSTEM

In the trophic stochastic logistical system, species are divided into compartments intended to reflect real food webs. A preliminary version of this system has three trophic compartments: producers, consumers, and decomposers, respectively. Individuals are chosen at random and those in the first compartment reproduce with a fixed probability, presumably deriving their energy from an outside source, such as plants absorbing energy from sunlight. When an individual is chosen from the second compartment, it "eats" an individual from the first compartment at random, reproducing in consequence. A variable that records the number of individuals consumed in either case is maintained for the third category; every time an individual is chosen from the decomposer compartment, it "eats" the remains of an individual that was removed from either of the first two compartments, decrementing the variable just defined. Just as producers come into being at random, so decomposers disappear at random. Admittedly, a crude reflection of energy flow in nature, this system also appears to produce the characteristic J shape, although that remains a purely visual assessment. Such a system might help to answer the question of how great a role trophism may play in the overall spectrum of stochastic influences.

10.5 CONCLUDING REMARKS

The research reported in this monograph has absorbed the better part of the author's last 20 years. I like to think that its appearance is timely in view of the ever-increasing need for a body of theory on which ecological science may find a footing. If the present criticisms of past methodologies are taken seriously, a certain procedural vacuum will result. In my opinion, the theory proposed here, with its sure-footed theoretical trail from the stochastic species hypothesis to the J distribution, will supply new content. With this monograph as a guide, there is no telling what new developments will emerge from the theory at hand. At present, the "community" of ecological theories is populated by many species, some abundant with followers, others rare. A new species has emerged, along with a question: Will it remain less abundant or will it meander as far as the delta-zone?

Appendix A: Mathematical Notes and Computer Tools

This appendix provides proofs of certain key results in the development of the hyperbolic theory. The first section includes proofs that the multispecies logistical system follows the J distribution with the parameter δ included. Reasons for the inclusion of the parameter ε are given in a second theorem. The second section consists of mathematical notes, and a final section describes the computer programs that have been used in this research program.

A.1 DERIVATION OF THE J DISTRIBUTION FOR STOCHASTIC SYSTEMS

The following theorem is referred to elsewhere in this monograph as the "equilibrium theorem." It provides a clear indication that the specific J-shape shared by virtually all samples is simply a consequence of the equilibrium process, which, from a mathematical (or statistical) point of view, is inevitable. The following result will be called the hyperbolic equilibrium theorem.

Theorem A.1

A stochastic system follows a distribution that is locally hyperbolic. ∎

Proof

Consider a stochastic system and a species of abundance k in it. At any given tick of the clock, each individual in the species has an equal probability p of increasing (a birth) or decreasing (a death) in abundance. Since the species has k individuals, the expected number of births must be pk. However, if there are F(k) species currently with abundance k, the expected number of births becomes pkF(k). Meanwhile, among the F(k + 1) species of abundance k + 1, the expected number of deaths is

$$p(k+1)F(k+1).$$

At equilibrium, the two quantities must be (statistically) equal, with as many species of abundance k increasing as the number of species of abundance k + 1 decreasing:

$$pkF(k) = p(k+1)F(k+1), \qquad (A.1)$$

which simplifies to

$$kF(k) = (k+1)F(k+1). \qquad (A.2)$$

This equation has a unique solution, namely, $F(k) = 1/k$, a hyperbola.

To see how equilibrium occurs, suppose that

$$kF(k) > (k+1)F(k+1),$$

so that $pkF(k) > p(k + 1)F(k + 1)$.

Comparing this inequality with Equation A.1, it is now more probable that a species of abundance k will increase than that a species of abundance k + 1 will increase. It is therefore more probable that, at the next iteration, the left-hand quantity will be smaller by unity than that $F(k + 1)$ will be smaller. Departures from hyperbolic values in F will therefore result in the increased probability of a compensatory increase or decrease in the corresponding F values. This completes the proof.   ∎

The theorem implies that the "mechanism" that produces the J distribution has nothing to do with specific ecological interactions or forms of ecological interaction, except insofar as they contribute collectively to the probabilities of increase and decrease. The hyperbolic shape just adduced is not the full story, of course, as revealed in the next theorem.

The dynamical behavior of a stochastic system has an interesting physical interpretation; if a thin cylinder of gas is heated at one end, the temperature gradient will ensure that molecules vibrate at different speeds, moving more quickly at the hot end and more slowly at the cool end. According to thermodynamic theory, the motions will be random. Molecules will nevertheless "prefer" to be at the cool end, since that yields a minimum overall energy for the system. If an imaginary wall is inserted anywhere along the tube, there would be a migration of gas molecules across the barrier in one direction that is matched by an approximately equal number of molecules passing through in the other direction, keeping pressure (density and energy of molecules) uniform across the barrier. The same reasoning that was used in the foregoing proof may be used here. The temperature decreases in linear fashion along the tube, with the slowest molecules tending to pile up at the cool end. The density of molecules in the cylinder accordingly increases in inverse linear fashion. The physics of gases is based on statistical mechanics, the science of collisions among molecules that comprise the gas.

Theorem A.2

A stochastic system has a hyperbolic global distribution translated downward by a specific amount.   ∎

Proof

Consider a given species of abundance k in a stochastic system as defined in Section 7.1.2. The probability p that an individual belonging to this species will die in the next time unit must be $p = k/N$, where N is the total number of individuals in the system. By definition, this is also the probability that a new individual will be born. It follows that the probability r that neither event will occur must be

$$r = 1 - 2k/N = 1 - vk,$$

where $v = 2/N$, a constant. Given that one of the F(k) species has just arrived at this size by the addition of a single individual, the appropriate expectation formula for the time it takes for the population size of this species to change once more is given by

$$E = \sum tr^t,$$

and the summation extends over successive time units $t = 0, 1, 2, \ldots, \infty$ on the clock. Expanding in terms of r yields,

$$E(k) = \sum t(1 - vk)^t.$$

Treating k as a continuous variable for the moment, one may now apply the differential operator D_k to the second factor of the summand above. The result will be used presently:

$$D_k(1 - vk)^t = (-vt)(1 - vk)^{t-1} \qquad (A.3)$$

The expression for E is next split into two factors as follows:

$$E(k) = \sum t(1 - vk)^{t-1}(1 - vk). \qquad (A.4)$$

Introducing two new factors v and $1/v$ that cancel, we obtain the following expression and then reinsert the differential operator, now applied to all k-terms of the series:

$$E(k) = (1/v) \sum (vt)(1 - vk)^{t-1}(1 - vk)$$
$$= (-1/v) \left[D_k \sum (1 - vk)^t \right] (1 - vk).$$

The operand is an infinite sum which may be put in closed form by using a standard convergence formula for the sum of a formal power series (see Section A.2.2).

$$\sum x^t = 1/(1 - x),$$

The summation converges to the right-hand side, provided that $x < 1$, which is true when $x = 1 - vk$. Applying the identity results in a simpler expression with a more tractable derivative,

$$E(k) = (-1/v)(1 - vk)D_k(1/vk),$$

so that

$$E(k) = (1 - vk)/k^2.$$

This represents the expected number of time units a species will retain the abundance k from its first arrival at k. However, species are arriving at abundance k and departing from it more or less continually. Returning to the first paragraph in this proof, if vk is the probability (per unit time) of a change in the abundance for the species in question, then the average time between arrivals must be $1/vk$.

Now if $E(k)$ is the average residence time for a species in the nth abundance category, and if $1/vk$ is the average interarrival time, then the number of species in that category, on average, must be

$$E(k)/(1/vk)$$

But the latter ratio must equal $F(k)$, which by definition is the number of species having abundance k. In other words,

$$F(k) = vk(1 - vk)/k^2$$
$$= v(1 - vk)/k \tag{A.5}$$

We conclude that the frequency function f must be proportional to K and so,

$$F(k) = c(1 - vk)/k,$$

where c is a constant. This completes the proof. ∎

A.1.1 Discrete Form and Emergence of the Second Parameter

The continuous form of the J function will obviously take the form of Equation A.5, generalized slightly from the rather specialized context in which it was derived:

$$f(x) = c(1/x - \delta),$$

where δ is a general parameter and c is a normalizing constant. Unfortunately, no such constant can exist because the area under the curve between 0 and unity is infinite. Reverting to the discrete form for the moment, the value $f(1)$ yields a new, second parameter ε that harmonizes the discrete and continuous forms of the distribution. We observe that there exists a real number ε such that

$$c\int_0^1 (1/(x+\varepsilon)-\delta)\,dx = f(1)$$

or $c(\ln((\varepsilon + 1)/\varepsilon) - \delta) = f(1)$.

This equation is easily solved for ε, although the resulting expression is neither elegant nor particularly useful. The conversion process has thus resulted in the appearance of a second parameter that also turns out, like the parameter δ, to be a translation, this time in the horizontal direction instead of the vertical one.

The two-parameter continuous density function, now denoted by f, becomes

$$f(x) = c(1/(x+\varepsilon)-\delta),$$

where c is a normalizing constant that brings the total density to unity,

$$c = (\ln(\Delta/\varepsilon)-1)^{-1},$$

where $\Delta = 1/\delta$.

The final result is a pure hyperbola translated by small amounts in both the horizontal (ε) and vertical (Δ) directions, respectively. The situation is readily grasped in terms of the diagram in Figure 1.2, reproduced here as Figure A.1.

The new axis system has its own origin, of course, and the parameters Δ and E (= $1/\varepsilon$) are displayed as x- and y-intercepts. The resulting simplicity and symmetry may be taken as hallmarks of a correct theory. The first parameter was forced upon us by the mathematics and the second factor was a necessary adjunct to harmonize the discrete and continuous forms of the J distribution.

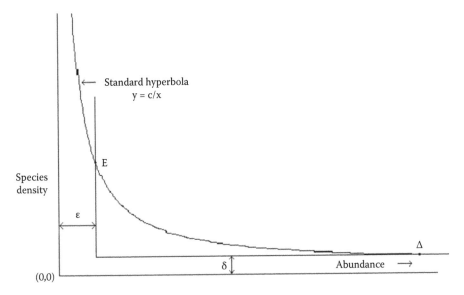

FIGURE A.1 The continuous J distribution in relation to the standard hyperbola.

A.2   USEFUL FORMULAS AND EQUATIONS

The following formulas are used in the mathematical developments that appear in this book. Each item in the following list includes a back-reference to the text that it supports.

A.2.1   INTEGRATION BY SUBSTITUTION YIELDS A FORMULA FOR THE MEAN

The integral here appears in the formula for the mean of the J distribution in Section 2.1.2.

$$c \int_0^{\Delta-\varepsilon} (x/(x+\varepsilon) - \delta x)\, dx$$

Integrals that do not correspond to standard integration formulas may sometimes be solved by a substitution of variables, as here. Substituting for the variable x, using $x' = x + \varepsilon$, we obtain

$$c \int_\varepsilon^{\Delta} \big((x'-\varepsilon)/x' - \delta(x'-\varepsilon)\big)\, dx' = c \int_\varepsilon^{\Delta} (1 - \varepsilon/x') - \delta x' + \delta\varepsilon)\, dx'$$

$$= c\Big[x' - \varepsilon \ln(x') - \delta x'^2/2 + \delta\varepsilon x')\Big]\Big|_\varepsilon^{\Delta}$$

$$= c(\Delta - \varepsilon - \varepsilon \ln(\Delta/\varepsilon) - \Delta/2 + \varepsilon + \delta\varepsilon^2/2 - \delta\varepsilon^2)$$

$$= c(\Delta/2 - \varepsilon \ln(\Delta/\varepsilon) + \varepsilon^2/2\Delta).$$

Except for the slight approximation built into c, this formula is exact. We note in passing that when $\varepsilon \ll \sqrt{\Delta}$, the last term can be neglected without altering the computed value of the integral significantly.

A.2.2   SUMMATION OF A POWER SERIES

The following result is used in Theorem A.2.

The infinite series Σx^k converges when $x < 1$, the sum being given in this case by the formula

$$\sum x^k = 1/(1-x).$$

To evaluate the expression $1/(1-x)$, simply use the long division technique learned in grade school:

1. The quantity $1 - x$ goes into 1 just 1 time, with a remainder of $1 - (1 - x) = x$;
2. The quantity $1 - x$ goes into x just x times with a remainder of $x - (x - x^2) = x^2$;
3. The quantity $1 - x$ goes into x^2 just x^2 times with a remainder of $x^2 - (x^2 - x^3) = x^3$, etc.

The result of the division process never terminates, and one is left with the infinite series $1 + x + x^2 + x^3 + \ldots$ At $x = 1$, the expression has the value ∞.

A.2.3 DERIVATION OF FORMULAS FOR THE MEAN AND VARIANCE OF THE J DISTRIBUTION (SECTION 2.1.2)

The mean value for the continuous version of the J distribution is defined as the integral

$$\mu = c \int_{\Sigma}^{\Delta - \Sigma} [x/(x+\varepsilon) - \delta x] dx$$

A change of variable such as $x' = x + \varepsilon$ rids us of the awkward $(x + \varepsilon)$ denominator:

$$= c \int_{e}^{\Delta} \left[(x' - \Sigma)/x' - \delta(x' - \Sigma) \right] dx$$

$$= c \int_{e}^{\Delta} \left[(1 - \Sigma/x') - \delta x' + \delta\Sigma) \right] dx$$

$$= c(x' - \Sigma \ln(x') - \delta x'^2/2 + \delta\Sigma x') \Big|_{\varepsilon}^{\Delta}$$

$$= c\left[\Delta - \Sigma \ln(\Delta) - \delta\Delta^2/2 + \delta\Sigma\Delta - \Sigma + \Sigma \ln)\Sigma) + \delta\Sigma^2/2 - \delta\Sigma^2 \right]$$

$$= c\left[\Delta - \Sigma \ln(\Delta) - \Delta/2 + \Sigma - \Sigma + \Sigma \ln(\Sigma) + \delta\Sigma^2/2 - \delta\Sigma^2 \right]$$

$$= c(\Delta/2 - \Sigma \ln(\Delta/\Sigma) - \delta\Sigma^2/2).$$

This completes the derivation of the formula for μ.

We turn now to the derivation of a formula for the variance of the J distribution, again using the continuous version. For the sake of simplicity, we will employ the formula

$$\text{Var}(x) = E[(x - \mu)^2] = E[x^2] - \mu^2$$

$$= c \int (x^2/(x+\varepsilon) - x^2\delta) dx - \mu^2.$$

Transform the variable: $x' = x + \varepsilon$, with x' running from ε to Δ.

$$
\begin{aligned}
\mathrm{Var}(x) &= c\int ((x'-\varepsilon)^2/x' - (x'-\varepsilon)^2\delta)\,dx' - \mu^2 \\
&= c\int (x'^2 - 2x'\varepsilon + \varepsilon^2)/x'\,dx' - c\delta\int (x'-\varepsilon)^2\,dx' - \mu^2 \\
&= c\int (x' - 2\varepsilon + \varepsilon^2/x')\,dx' - c\delta(x'-\varepsilon)^3/3 - \mu^2 \\
&= c(x'^2/2 - 2x'\varepsilon + \varepsilon^2\ln(x') - c\delta(x'-\varepsilon)^3/3 - \mu^2.
\end{aligned}
$$

Evaluating at the integration limits yields

$$
\begin{aligned}
&= c(\Delta^2/2 - 2\Delta\varepsilon + \varepsilon^2(\ln(\Delta) - (\varepsilon^2/2 - 2\varepsilon^2 + \varepsilon^2\ln(\varepsilon)) - c\delta(\Delta-\varepsilon)^3/3) - \mu^2 \\
&= c[(\Delta^2 - \varepsilon^2)/2 - 2(\Delta-\varepsilon)\varepsilon + \varepsilon^2(\ln(\Delta) - \ln(\varepsilon)) - \delta(\Delta-\varepsilon)^3/3] - \mu^2 \\
&= c[(\Delta^2 - \varepsilon^2)/2 - 2(\Delta-\varepsilon)\varepsilon + \varepsilon^2(\ln(\Delta/\varepsilon)) - \delta(\Delta-\varepsilon)^3/3] - \mu^2,
\end{aligned}
$$

with $\ln(\Delta/\varepsilon) = 1/c + 1$ we have,

$$
\begin{aligned}
\mathrm{Var}(x) &= c[(\Delta^2 - \varepsilon^2)/2 - 2(\Delta-\varepsilon)\varepsilon + \varepsilon^2(1/c+1) - \delta(\Delta-\varepsilon)^3/3] - \mu^2 \\
&= c[(\Delta^2 - \varepsilon^2)/2 - 2(\Delta-\varepsilon)\varepsilon + \varepsilon^2 - \delta(\Delta-\varepsilon)^3/3] + \varepsilon^2 - \mu^2 \\
&= c[(\Delta^2 - \varepsilon^2)/2 - 2\Delta\varepsilon + 3\varepsilon^2 - \delta(\Delta-\varepsilon)^3/3] + \varepsilon^2 - \mu^2 \\
&= c[(\Delta^2/2 - 2\Delta\varepsilon + (2.5)\varepsilon^2 - \delta(\Delta-\varepsilon)^3/3] + \varepsilon^2 - \mu^2.
\end{aligned}
$$

A.2.4   DERIVATION OF THE CANONICAL FORMULA (SECTION 2.2.2)

The discrete version of the logistic-J distribution described in Section 2.2.1 summarizes the expected numbers of species per abundance category. Another description of the distribution lists the expected abundances of the species themselves. The canonical abundance a_k of the kth species (in abundance rank) is described in Section 2.2.2 and is readily found via the integral equation

$$
\mathrm{Rc}\int_0^{a_k} (1/(x+\varepsilon) - \delta)\,dx = k,
$$

By integrating over the J distribution up to the expected kth abundance a, then solving the resulting equation, we will have "captured" k species and know the abundance of the kth most numerous species.

Thus, the expected abundance a_1 of the first species may be found by solving for a in

$$Rc(\ln(x+\epsilon)-\delta x)\Big|_0^a = 1$$

The solution is easy:

$$Rc(\ln((a+\epsilon)/\epsilon)-\delta a) = 1.$$

Letting $T = 1/Rc$, we have

$$\ln((a+\epsilon)/\epsilon)-\delta a = T.$$

This equation can be solved for a only by iterative methods. Call the solution a_1.

The position of the next canonical abundance, a_2, is the solution for a in the equation

$$Rc\int_0^a (1/(x+\epsilon)-\delta)\,dx = 2.$$

In this case,

$$\ln((a+\epsilon)/\epsilon)-\delta a = 2T.$$

The general equation is clearly

$$\ln((a+\epsilon)/\epsilon)-\delta a = kT, \qquad (A.6)$$

where k indicates the kth species in the sequence.

If we remove the term δa (being negligible for small values of k), the equation becomes directly solvable, yielding

$$(a+\epsilon)/\epsilon = \exp(kT)$$

or

$$a_k = \epsilon(\exp(kT))-\epsilon. \qquad (A.7)$$

Loss of the term δa has little effect on the solution, since the factor $\epsilon^{-\delta c}$ that would otherwise appear in Equation A.7 is very close to unity when δc is small (typically < 0.01).

A.2.5   PROOF OF THE RANDOM HIERARCHY

The random hierarchy is defined in Section 3.1 and proved here to be real.

Theorem A.3

If a procedure is k-random, then it is also (k − 1)-random, for k > 1.   ∎

Proof

Let P be an effectively k-random procedure. Then every sequence of length k has an equal probability of being generated by P. This probability must be $p = 2^{-k}$. Let S be an arbitrary sequence of length k − 1 and suppose that it will appear with probability q. The sequences (S, 0) and (S, 1) have length k and the probability that *either* sequence will appear must be q = 2p, using the additive property of independent events. But this implies that $q = 2^{-(k-1)}$.   ∎

A.2.6   RANDOM TIME SERIES FOLLOWS THE EXPONENTIAL DISTRIBUTION

The negative exponential formula for time series is discussed in Section 7.1.1 and proved here. Let A be a time series that satisfies the two following conditions:

1. A long-term average of λ time units between events prevails.
2. Events are statistically independent; the probability of an event at any point in time depends only on the time since the last event.

Theorem A.4

The distribution of interevent times in A follows the negative exponential distribution.   ∎

Proof

Consider a small time increment dt and note that the probability of an arrival during this small period is dt/λ. As seen already in Section 7.1.1, this is only an approximation, but very accurate for small intervals like dt. The probability that no event will occur during this interval is

$$1 - (dt/\lambda).$$

Let f(t) be the probability that the next event will occur at least t time units after the previous one. The expression f(t + dt) represents the probability that the next event will occur at least t + dt time units after the previous one. The latter probability compounds two independent possibilities, namely, that (a) an event does not occur during the first t time units and (b) an event does not occur in the subsequent period of duration dt.

Since the next event must occur still later than t + dt time units, it follows that

$$f(t+dt) = f(t)(1-dt/\lambda),$$

by the product rule for independent probabilities. By the definition of the differential,

$$df/dt = (f(t+dt)-f(t))/dt$$
$$= (f(t)(1-dt/\lambda)-f(t))/dt$$
$$= (f(t)/dt)(1-dt/\lambda-1)$$
$$= -f(t)/\lambda,$$

taking the limit as λ approaches 0, the result is immediate since dt has vanished.

$$df/dt = \lim_{dt \to 0}(-f(t)/\lambda) = -f(t)/\lambda$$

Using the boundary condition f(0) = 1, the function f can be recognized as the exponential function because of the unique property of the exponential function that it equals its own derivative.

$$f(t) = e^{-t/\lambda}$$

Now f(t)dt represents the probability of an event between time t and time t + dt. The density function is simply this probability divided by dt. We now have two independent probabilities:

$e^{-t/\lambda}$, the probability of no event before t time units and
dt/λ, the probability of an event within dt time units.

Applying the product rule once more reveals the form of the density function:

$$(e^{-t/\lambda}\, dt/\lambda)/dt = 1/\lambda e^{-t/\lambda},$$

as required.   ∎

A.2.7   Log-Series Distribution (Section 7.1)

The general term of the log-series distribution, as mentioned in Section 8.3.2, is

$$F(k) = \alpha c^k /k,$$

where α is a constant defined by

$$\alpha = N(1-c)/c,$$

with N being the total number of individuals in the sample and c being a free parameter that normally takes values that are less than unity, but close to it.

Since the sum $\Sigma F(k)$ yields the number of species, R, and since the series sums to $\alpha(-\ln(1-c))$, one has

$$R = \alpha(-\ln(1-c)).$$

It now follows that c may be found by solving the equation

$$(1-c)(-\ln(1-c)/c = R/N = 1/\mu).$$

The probability density function for the log-series distribution is obtained simply by substituting for the mean abundance μ in a specific instance, then solving for the constant c.

$$F(k) = \alpha c^k / k$$

Thus, the log-series distribution is discrete and has just one parameter, namely, c, the pseudo-parameter α being actually a joint function of c and μ. The total number of degrees of freedom for a chi-square test of an n-category histogram with the log-series distribution would therefore be $n - 1$, whereas the J distribution has $n - 2$ degrees of freedom. Consequently, no optimization is possible for the log-series distribution, since α and c are determined completely by μ.

A.2.8   TRANSFER EQUATIONS

The transfer equations are a key tool in the meta-study, as described in Section 8.3.2. They are also mentioned in Sections 4.4 and 5.1. The equations permit us to express one set of parameters and/or constants in terms of another. The application developed here enables us to infer a value for epsilon from values for μ and F_1, the number of species of abundance 1, as well as delta, the maximum expected abundance.

The starting point for developing the transfer equation is our formula (Section 2.1.2) for the mean.

$$\mu = c((\Delta - \varepsilon)(\Delta + \varepsilon)/2\Delta) - \varepsilon \ln(\Delta/\Sigma) \tag{A.8}$$

Since $1/c = (\ln(\Delta/\varepsilon) - 1 + \varepsilon/\Delta)$, we may write

$$\ln(\Delta/\varepsilon) = 1/c + 1 - \varepsilon/\Delta$$

and substitute this value into Equation A.8 to obtain

$$\mu = c((\Delta - \varepsilon)(\Delta + \varepsilon)/2\Delta - \varepsilon(1/c + 1 - \varepsilon/\Delta))$$
$$= c((\Delta - \varepsilon)(\Delta + \varepsilon)/2\Delta - \varepsilon(1 - \varepsilon/\Delta)) - \varepsilon.$$

So, $\mu + \varepsilon = c((\Delta - \varepsilon)(\Delta + \varepsilon)/2\Delta - \varepsilon + \varepsilon^2/\Delta)$.

Thus, we have eliminated the logarithmic term and may now write

$$c = (\mu + \epsilon)((\Delta - \epsilon)(\Delta + \epsilon)/2\Delta - \epsilon + \epsilon^2/\Delta)^{-1}. \qquad (A.9)$$

I will call this equation *system 1*.

Consider now the height of the first abundance category in the discrete form of the logistic-J pdf. The height (number of species) in the first abundance category is given by the integral,

$$F(1) = Rc \int_0^1 (1/(x + \epsilon) - \delta)\,dx$$

$$= Rc(\ln(x + \epsilon) - \delta x)\Big|_0^1.$$

Simplifying F(1) notationally to F_1, we may write

$$F_1 = Rc(\ln((1 + \epsilon)/\epsilon) - 1/\Delta)$$

or

$$c = (F_1/R)(\ln((1 + \epsilon)/\epsilon) - 1/\Delta)^{-1}. \qquad (A.10)$$

The two expressions for c (Equations A.9 and A.10) may now be equated, yielding

$$(\mu + \epsilon)((\Delta - \epsilon)(\Delta + \epsilon)/2\Delta - \epsilon + \epsilon^2/\Delta)^{-1} = (F_1/R)(\ln((1 + \epsilon)/\epsilon) - 1/\Delta)^{-1}.$$

Let $F_1/R = T$ and rearrange the terms of the previous equation to yield the following:

$$(\mu + \epsilon)(\ln((1 + \epsilon)/\epsilon) - 1/\Delta) = T((\Delta - \epsilon)(\Delta + \epsilon)/2\Delta - \epsilon + \epsilon^2/\Delta).$$

Multiplying through by Δ and simplifying further, one obtains

$$(\mu + \epsilon)(\Delta \ln((1 + \epsilon)/\epsilon) - 1) = T((\Delta - \epsilon)(\Delta + \epsilon)/2 - \epsilon\Delta + \epsilon^2).$$

The next step in the development of the transfer equation is to expand the previous expressions then to collect like terms in powers of Δ:

$$(\mu + \epsilon)\Delta \ln((1 + \epsilon)/\epsilon) - (\mu + \epsilon) = T((\Delta^2 - \epsilon\Delta + \epsilon\Delta - \epsilon^2)/2 - \epsilon\Delta + \epsilon^2),$$

$$(\mu + \epsilon)\Delta \ln((1 + \epsilon)/\epsilon) - (\mu + \epsilon) = T\Delta^2/2 - T\epsilon^2/2 - T\epsilon\Delta + T\epsilon^2,$$

and

$$(T/2)\Delta^2 - (T\varepsilon + (\mu + \varepsilon)\ln((1+\varepsilon)/\varepsilon))\Delta + (\mu + \varepsilon + T\varepsilon^2/2) = 0.$$

Now let

$$A = T/2,$$

$$B = T\varepsilon + (\mu + \varepsilon)\ln((1+\varepsilon)/\varepsilon),$$

$$C = \mu + \varepsilon + T\varepsilon^2/2$$

and note that the equation $A\Delta^2 - B\Delta + C = 0$ is a quadratic in Δ and therefore has a solution of the form,

$$\Delta = \left(B \pm \sqrt{B^2 - 4AC}\right)/2A \qquad\qquad (A.11)$$

Note also that Δ can be computed in one step, via Equation A.11, since none of the three expressions involving A, B, and C contain Δ. I will call Equation A.11 *system 2*.

Solutions of the transfer equations may be found by using Newton's method, as embodied in the program SolveIt (described in Section A.3.4). The underlying algorithm proceeds iteratively by choosing a starting value for ε, then substituting the values on hand for R, F_1, and μ into system 2. This produces a value for Δ that is then substituted, along with the values for ε and μ, into system 1. The resulting value for ε is then recycled back into system 2, where the process begins anew. Over a number of iterations, the value of ε (as well as that of Δ) converges to a stable value. These values are the ones implied (in the mathematical sense) by the observed values of R, F_1, and μ.

A more general form of the transfer equation can be developed from the foregoing systems by replacing the integration limit 1 in system 2 by the parameter a, which can be any abundance, whether integral or fractional. This yields the equation

$$c = T(\ln((a+\varepsilon)/\varepsilon) - a/\Delta)^{-1}$$

on the way to a more general version of system 2, whereas system 1 remains the same, with $T = F_a/R$.

$$A = T/2$$
$$B = T\varepsilon + (\mu + \varepsilon)\ln((a+\varepsilon)/\varepsilon)$$
$$C = a(\mu + \varepsilon) + T\varepsilon^2/2$$

A.2.9 SAMPLING THE UNIVOLTINE DISTRIBUTION

The univoltine (Latin—one leap) distribution is used as a rich source of counter-examples in Chapters 3 and 5, as well as playing an important role in the development of accumulation curves in Section 6.2. In this note, we complete the argument made in Section 6.2 to derive a recurrence relation in the solution of the accumulation problem for species as a function of sample size. The example will involve just part of a community, a single column at abundance 8 and 10 species of that abundance: U[8] × 10.

Let R(k) represent the expected number of species to have appeared in the sample at the time the kth individual is observed. Given a sample of size k, what is the probability of a new species appearing at the next observation? The probability that the next individual will belong to a new species will depend on the number of individuals belonging to the set of $10 - R(k)$ species that have not yet been sampled. The number is, of course, $8 \times (10 - R(k))$ and the associated probability must be

$$8(10 - R(k))/80 \quad \text{or} \quad 1 - 0.1R(k).$$

The expected number of new species at the (k + 1)st drawing is simply this probability. When added to R(k), we obtain the following recurrence relation:

$$R(k+1) = 0.9R(k) + 1.$$

In such relations, the value of a function at k is stated as a function of previous values, often more than one, such as $k - 1$, $k - 2$, and so on. It would be convenient to have a closed-form version of the formula that would be easier to calculate. Before solving the recurrence relation, however, I will show how the recurrence may be used to generate values of R(k) for any value of k. Starting at $R(1) = 1.0$, for example, the relation tells me that $R(2) = 1.9$ and applying the relation once again, I obtain $R(3) = 2.71$. Table A.1 gives the first 20 values of the function R as calculated by this method.

A simple empirical test reveals the accuracy of the formula when compared with the results of a simulated sampling. Table A.2 shows the comparison at a few evenly spaced points.

The match between empirical and theoretical values gives us confidence that no mistake was made in deriving the recurrence relation. To obtain a closed form solution, we must solve the recurrence.

$$R(k) = 0.9R(k-1) + 1 \tag{A.12}$$

One way to solve recurrence relations in general is to back-substitute, replacing $R(k - 1)$, for example, by

$$0.9R(k-2) + 1.0$$

TABLE A.1
Expected Number R(k) of Species
as a Function of Sample Size

k	R(k)
1	1.000
2	1.900
3	2.710
4	3.349
5	4.095
6	4.685
7	5.216
8	5.695
9	6.126
10	6.513
11	6.862
12	7.176
13	7.458
14	7.712
15	7.941
16	8.147
17	8.332
18	8.499
19	8.649
20	8.784

TABLE A.2
Comparison of Theoretical R and Empirical S Sample Riches

k	4	8	12	16	20
R(k)	3.35	5.69	7.18	8.15	8.78
S(k)	3.45	5.50	7.00	8.20	8.80

in Equation A.12, then continuing this process down to the last replacement by R(1). The general result for the example follows:

$$R(k) = \sum_{i=0}^{k} 0.9^i. \tag{A.13}$$

This expression for the formula is slightly more useful than the earlier recurrence relation, but the sum cannot be simplified. In the limit, however, we may apply the well-known formula for an infinite geometric series (Appendix A.2.2) to this component of the formula, yielding

$$\text{Lim } R(k) = 0 + 1.0/(1 - 0.9) = 10.0, \text{ as expected.}$$
$$k \to \infty$$

Can Equation A.13 be generalized to any univoltine distribution? Obviously it can. We cut to the chase by replacing the constant 0.9 in Equation A.13 by $(1 - 1/R)$:

$$R(k) = \sum_{i=0}^{k} (1 - 1/R)^i.$$

This completes the solution of the accumulation curve for the univoltine distribution with replacement sampling.

A.3   COMPUTER RESEARCH TOOLS

It seems fair to say that no successful theory of population or community dynamics is possible without the aid of computer software that is capable of simulating the sampling process accurately in a statistical sense or emulating dynamical processes in 1–1 fashion. Armchair insights will often founder on the rocks of a decisive computer test.

Each program is listed by its filename, followed by a description of its operation. Several of the key programs used in this research program will be found in functional form in a graphic user interface on the author's website at www.csd.uwo.ca/~akd/.

A.3.1   Systems Emulators

The Multispecies logistic system: This program comes in two versions and is certainly open to development in other versions, such as the trophic compartment system as described briefly in Section 10.4.4.

The Basic MLS program: Informally named "Scramble," this program takes an initial list of species abundances as input and then enters a continuing cycle of (a) selecting an individual at random, (b) deciding with equal probability on a birth or death event for that individual, (c) adding or subtracting 1 from the corresponding population of that species according to the decision in step b. An extirpation switch, when turned "off," causes any death event for a population of 1 to be cancelled. When the user presses an "end" key, the process stops and the resulting biodiversity vector is printed out or stored as a file.

The Weakly Stochastic MLS program, also known as StoComm: The program is similar to the Basic MLS program, except for a more complicated loop structure internally. Having selected an individual at random, it consults a table of what might be called primary probabilities that determine which entry p in a table of secondary probabilities will be accessed next. That probability is then applied directly to the birth/death decision process. The randomly chosen individual will then become two individuals or none, both with probability p.

A.3.2   SAMPLING AND ESTIMATION PROGRAMS

CommRich: The program CommRich is a richness-estimation program based directly on hyperbolic theory. Using either the raw sample histogram or the best-fit version of it, this program computes the expected sample from a community and compares it with the sample data via the least squares measure. To do this, it needs a sampling intensity estimate from the user. The program structure is reentrant, allowing the user to refine parameter estimates, converging on a final value for richness R.

SampleSim: This sampling program takes intensity r, along with parameters of the community $J[\varepsilon, \Delta] \times R$, as input. It then "samples" this community at intensity r, producing a sample of size rN, where N is the total population to be sampled.

A.3.3   STATISTICAL PROGRAMS

ChiSquare: This program takes an empirical (field) distribution as input and compares it with either the logistic-J or log-series distributions, as specified by the user. It uses the standard rule-of-five that, until a category has accumulated a total of at least 5.0 points of the theoretical distribution, the category is extended until the criterion is met. The program outputs the chi-square score as well as the number of degrees of freedom used by the fitting process.

BestFit: This program employs the chi-square test in reentrant fashion, allowing the user to adjust parameters as the computation proceeds. The user must find the combination of parameter values that produces the minimum chi square score.

ScanFit: In this version of the score minimization program, the user specifies a grid of test points in parameter space. The program computes the score for each combination, ultimately reporting on the combination that produced the minimum score.

KSTest: This program calculates the Kolmogorov-Smirnov statistic for an empirical distribution specified by the user against a particular J distribution, also specified by the user.

A.3.4   SPECIAL UTILITY PROGRAMS

SolveIt1: This program solves the logistic equation iteratively, starting with an initial guess by the user. This guess may be critical in the sense that if it is too large or too small, the iterative process may diverge. Other inputs include richness R, mean μ, category width a (usually unity), and F_a, the number of lowest-abundance species. The program outputs the value for ε that solves the system, as well as values for Δ, δ, and the coefficient c.

Solvelt2: This program solves another set of transfer equations that carry ε as an implicit function of μ and Δ instead of μ and F_a.

Canon: This program calculates the canonical sequence for the specified distribution $J[\varepsilon,\Delta] \times R$. Presently, this program operates in interactive mode. It can be rewritten to find each canonical abundance without human intervention.

HGen: This program generates theoretical values of a given logistic-J distribution with parameters, ε, Δ, and richness R. To do this, it uses the formula for the J distribution pdf, then multiplies it by R.

Overlap: This program computes the expected overlap of two samples based on an empirically derived community, $J[\varepsilon,\Delta] \times R$.

Interpol: Given a chi-square score at n degrees of freedom, Interpol finds the equivalent score at 10 degrees of freedom by using linear interpolation on the entries of a chi-square table.

Appendix B: Results of the Meta-Study for the J Distribution

The following tables display the results of the meta-study as they pertain to the logistic-J distribution alone. Beside each biosurvey index number is the number R of species in the corresponding sample, the number F_a of species having minimum abundance, the average abundance m, and parameter values. The parameter Δ' is the maximum abundance, Δ is the predicted logistic limit, and ε is the predicted epsilon-value. The last three columns contain the chi-square score achieved by the F_a-μ fitting process, the same score normalized to 10 degrees of freedom, and the ratio of actual to predicted maximum abundance, expressed as a percentage.

#	R	F_a	μ	Δ'	Δ	ε	F_a-μ Score	Normalized	Δ'/Δ
1	222	4.8	9.338	52	70.0	0.907	18.55/18	10.25	74.3
2	79	14	6.901	25	37.2	2.279	5.22/8	6.91	67.2
3	31	6	5.548	32	27.6	2.478	4.64/3	13.40	115.9
4	41	9	29.44	297	268.7	1.209	0.281/4	2.47	110.5
5	87	9	66.88	704	571.8	3.844	5.71/12	4.19	123.1

#	R	F_a	μ	Δ'	Δ	ε	F_a-μ Score	Normalized	Δ'/Δ
6	39	4	622.0	6053.7	5769.4	26.11	6.695/5	12.59	104.9
7	38	7	5.470	16	25.8	3.048	1.490/4	7.36	62.0
8	36	12	25.47	173	255.9	0.7265	0.5395/3	4.69	67.6
9	44	12	443.0	5797	6527.9	1.559	1.039/5	3.71	88.8
10	79	23	4.259	27	25.5	0.8293	7.633/6	12.28	101.9

#	R	F_a	μ	Δ'	Δ	ε	F_a-μ Score	Normalized	Δ'/Δ
11	33	7	47.79	285	397.2	3.083	0.4434/3	4.30	71.8
12	49	16	3.690	44.4	42.1	0.057	0.4424/5	2.47	105.5
13	183	71	7.344	37	55.5	0.693	4.0441/9	5.16	66.7
14	41	35	35.68	406	362.5	0.9667	40.066/29	16.64	112.0
15	36	10	73.17	656	789.5	1.4885	5.5238/3	14.82	83.1

#	R	F_a	μ	Δ'	Δ	ε	F_a-μ Score	Normalized	Δ'/Δ
16	30	7	535.5	4716	8846.4	0.8545	4.8641/3	13.76	53.3
17	86	26	7.880	74.9	70.4	0.3763	7.7786/8	9.83	106.4
18	40	17	8.812	70	113.6	0.0701	0.0701/3	7.91	61.6
19	66	26	8.764	67.5	104.2	0.1090	5.8973/5	11.51	64.8
20	41	14	59.29	768	1051.4	0.0550	3.3584/4	9.35	73.0

#	R	F_a	μ	Δ'	Δ	ε	F_a-μ Score	Normalized	Δ'/Δ
21	113	24	39.22	1196	492.1	0.3639	23.456/14	17.22	240.0
22	78	18	69.04	839	927.7	0.4317	16.746/10	16.75	90.4
23	70	8	30.07	208	266.8	1.5090	4.2951/8	5.82	78.3
24	54	11	187.1	3829	2835.7	0.5959	6.2588/7	9.21	135.0
25	130	23	165.0	2764	2297.7	0.8304	23.959/17	15.34	120.3

#	R	F_a	μ	Δ'	Δ	ε	F_a-μ Score	Normalized	Δ'/Δ
26	222	31	5.809	37.9	68.8	0.0740	7.1344/5	13.22	55.1
27	140	70	6.971	77	100.7	0.0280	9.6287/9	10.70	76.5
28	295	80	6.763	48	51.1	0.6570	8.7739/18	3.60	93.9
29	63	17	9.238	63	81.1	0.4756	2.8055/6	5.77	77.7
30	620	118	14.57	194	122.9	0.8824	38.357/41	3.91	157.9

#	R	F_a	μ	Δ'	Δ	ε	F_a-μ Score	Normalized	Δ'/Δ
31	54	15	73.59	2073	1234.7	0.1097	11.3769/6	16.86	50.9
32	203	57	54.82	2324	725.9	0.3740	37.263/20	22.97	320.2
33	39	10	90.33	713	995.1	1.6595	3.1228/4	9.92	71.7
34	143	55	14.92	184.1	133.7	0.7053	13.2556/10	14.26	137.7
35	31	9	40.87	168	478.2	0.5525	4.4388/3	13.08	35.1

#	R	F_a	μ	Δ'	Δ	ε	F_a-μ Score	Normalized	Δ'/Δ
36	35	12	54.33	464	616.9	0.8571	2.3726/3	9.35	75.2
37	45	9	14.02	192	120.3	0.7942	11.505/5	18.84	159.6
38	42	14	1.492	18.54	18.5	0.0148	9.3077/4	18.06	100.0
39	71	11	365.2	5175	4947.5	2.1746	18.6046/10	18.60	104.6
40	195	31	3.644	20.0	25.3	0.2877	27.078/15	20.13	79.1

#	R	F_a	μ	Δ'	Δ	ε	F_a-μ Score	Normalized	Δ'/Δ
41	108	30	41.27	508.6	604.0	0.1518	9.5463/12	9.55	84.2
42	56	15	370.5	9418	5170.0	1.8414	8.2422/6	13.04	182.2
43	40	7	59.56	578.4	626.7	1.3672	0.7074/5	2.94	92.3
44	67	31	4.973	54.8	55.9	0.0828	7.6630/4	15.82	98.0
45	43	11	26.21	229	324.1	0.2634	0.7369/4	4.01	70.7

#	R	F_a	μ	Δ'	Δ	ε	F_a-μ Score	Normalized	Δ'/Δ
46	34	8	445.4	4433	4992.3	7.5043	0.2439/3	3.39	88.8
47	53	26	1561	54273	42918.5	0.0170	3.4966/4	9.49	126.5
48	105	22	53.12	1320	789.4	0.1768	12.8128/10	12.81	167.2
49	100	23	30.17	276	290.9	1.0361	17.6269/11	16.33	94.8
50	48	14	78.58	1058	1169.2	0.3596	5.1224/5	10.43	90.5

#	R	F_a	μ	Δ'	Δ	ε	F_a-μ Score	Normalized	Δ'/Δ
51	46	18	21.89	337	346.3	0.0476	0.9643/4	4.58	97.3
52	82	24	7.671	157	65.9	0.4329	14.6702/7	18.98	238.2
53	39	14	126.6	1625	2195.3	0.1400	4.0202/3	12.39	74.0
54	53	22	10.66	185.5	208.2	0.0044	2.8309/4	8.39	89.1
55	116	21	105.5	1743	1330.4	0.9517	13.9705/15	9.05	131.0

#	R	F_a	μ	Δ'	Δ	ε	F_a-μ Score	Normalized	Δ'/Δ
56	36	9	120.1	1115	1427.4	1.4936	6.7759/4	14.95	78.1
57	32	11	3.125	31.7	28.0	0.1468	1.8438/2	10.40	113.2
58	40	7	130.5	1009.9	998.2	11.726	1.3529/4	5.50	101.2
59	181	49	10.96	167	104.3	0.3989	14.787/16	8.92	160.1
60	44	15	16.18	99	162.4	0.4633	2.1758/4	7.20	61.0

#	R	F_a	μ	Δ'	Δ	ε	F_a-μ Score	Normalized	Δ'/Δ
61	36	12	5.573	35.9	28.0	2.4197	0.5312/2	6.51	94.5
62	81	15	78.57	1067	923.8	1.0356	9.8996/10	9.90	115.5
63	41	9	9.610	61	73.8	0.8468	4.9002/4	11.78	82.7
64	35	7	140.9	2716	1558.2	2.5421	3.3237/4	9.28	174.3
65	43	8	1341	20,190	21,119	3.0186	2.7733/5	6.83	95.6

#	R	F_a	μ	Δ′	Δ	ε	F_a-μ Score	Normalized	Δ′/Δ
66	38	11	17.42	335	206.2	0.2274	3.3063/4	9.25	162.5
67	36	8	20.72	146	166.5	1.5305	0.1229/3	2.61	87.7
68	75	18	48.84	467	570.0	0.6692	10.1624/8	11.26	81.9
69	73	25	4.880	34	38.2	0.4012	3.1293/6	6.25	89.0
70	39	9	7.950	26	57.5	0.8822	4.5086/4	11.16	45.1

#	R	F_a	μ	Δ′	Δ	ε	F_a-μ Score	Normalized	Δ′/Δ
71	34	7	15.97	78	81.6	6.4782	4.1511/3	12.62	95.6
72	81	18	27.88	209	324.7	0.3863	10.1709/9	11.27	64.4
73	34	6	103.4	1029	1144.2	1.8620	2.5627/4	7.90	89.9
74	79	17	25.51	162	248.1	0.8434	0.3078/5	1.61	65.3
75	41	10	799.4	6285	12,697	1.6910	2.9621/5	7.18	49.5

#	R	F_a	μ	Δ′	Δ	ε	F_a-μ Score	Normalized	Δ′/Δ
76	50	12	132.5	2084	2308.2	0.1414	11.4995/6	17.01	90.4
77	36	10	48.60	309	739.5	0.1382	2.6965/3	9.95	41.8
78	36	10	2.778	15.2	36.9	0.0193	2.4169/4	7.62	41.2
79	60	23	7.033	61	74.3	0.1585	12.2579/5	19.77	82.1
80	51	10	89.20	761	983.1	1.6351	4.8208/6	8.62	77.4

#	R	F_a	μ	Δ′	Δ	ε	F_a-μ Score	Normalized	Δ′/Δ
81	32	11	91.16	1195	1795.2	0.0351	4.0404/4	10.42	66.2
82	37	15	21.49	3470	2304.9	0.0387	8.5866/3	19.40	150.5
83	36	14	1.519	11.7	21.54	0.0068	1.8215/3	8.11	54.4
84	46	15	74.59	1247	1097.9	0.1729	4.7117/4	11.48	113.6
85	31	26	6.405	146	168.9	0.0001	0.6349/1	10.01	86.4

#	R	F_a	μ	Δ′	Δ	ε	F_a-μ Score	Normalized	Δ′/Δ
86	50	15	68.74	675	1006.5	0.2521	8.5065/5	15.04	67.1
87	121	51	4.339	38	40.2	0.1773	13.4560/8	16.09	94.5
88	31	14	39.79	609	445.6	0.6731	6.8828/6	11.32	136.7
89	62	14	6.694	37	43.2	1.1432	7.6460/6	12.30	85.6
90	51	10	14.78	161	128.1	0.8033	5.0197/6	8.89	125.7

#	R	F_a	μ	Δ′	Δ	ε	F_a-μ Score	Normalized	Δ′/Δ
91	61	13	7.443	24	33.0	1.1967	7.622/6	6.75	72.7
92	29	8	3.459	15.4	39.4	0.0540	0.5681/2	6.70	39.1
93	58	26	1.879	18.5	22.9	0.0203	6.6478/5	12.58	80.8
94	47	15	629.8	7912	11003	0.6575	4.4453/5	9.47	71.9
95	38	9	39.95	262	330.2	2.6381	8.8212/4	17.50	79.3

#	R	F_a	μ	Δ′	Δ	ε	F_a-μ Score	Normalized	Δ′/Δ
96	118	26	35.57	780	428.3	0.4136	20.372/15	14.46	182.1
97	52	13	11.92	126	110.6	0.4856	5.3194/5	10.71	113.9
98	38	10	127.6	1279	2005.9	0.2908	2.6119/4	7.99	63.8
99	35	17	24.63	249	414.6	0.0342	11.0463/2	22.63	60.1
100	50	18	269.0	2830	4926.7	0.1924	3.0045/5	5.83	57.4

#	R	F_a	μ	Δ′	Δ	ε	F_a-μ Score	Normalized	Δ′/Δ
101	32	9	55.91	260	549.5	1.7626	0.5664/3	4.35	47.3
102	31	10	19.19	124	197.3	0.4917	1.9449/2	10.63	62.8
103	34	10	6.88	32	56.2	0.4758	2.4368/3	9.47	56.9
104	49	10	16.78	148	156.7	0.6635	5.9845/6	11.64	94.4
105	30	11	213.4	2681	3176.0	0.4758	4.1667/2	15.16	84.4

#	R	F_a	μ	Δ′	Δ	ε	F_a-μ Score	Normalized	Δ′/Δ
106	61	19	7.23	56	41.7	1.8601	6.3239/5	12.12	134.3
107	34	6	172.5	1080	1067.7	34.282	0.9082/3	5.83	101.1
108	170	24	109.5	6379	1495.2	0.6218	27.128/23	12.54	426.6
109	37	15	7.13	78	80.4	0.1169	2.7795/3	10.10	97.0
110	51	9	57.4	1057	732.2	0.4872	3.2564/6	6.64	144.4

#	R	F_a	μ	Δ′	Δ	ε	F_a-μ Score	Normalized	Δ′/Δ
111	39	10	42.05	441	591.5	0.1990	3.1687/4	9.00	74.6
112	29	5	0.242	2.30	1.77	0.0259	0.1989/3	2.41	129.9
113	53	16	0.051	1.64	0.68	0.0003	7.7344/5	14.01	241.2
114	46	20	73.35	740	1446.2	0.0279	5.0963/4	12.09	51.2
115	33	8	9.757	37	64.4	2.8463	1.5775/3	7.56	57.7

#	R	F_a	μ	Δ′	Δ	ε	F_a-μ Score	Normalized	Δ′/Δ
116	31	9	3.223	24.6	31.9	0.0982	0.9271/3	5.89	77.1
117	31	5	178.1	2836	2435.5	1.0024	7.0013/3	17.10	116.4
118	29	8	7.32	47.0	58.5	0.5535	1.3449/2	9.20	80.3
119	31	7	608.5	5879	5847.2	0.9415	4.3255/3	12.90	100.5
120	35	6	3.609	18.6	19.5	1.1770	4.0507/4	10.44	95.4

#	R	F_a	μ	Δ′	Δ	ε	F_a-μ Score	Normalized	Δ′/Δ
121	123	39	3.665	49.2	56.4	0.0096	10.445/13	7.72	87.2
122	64	17	12.14	130	118.3	0.3972	3.0635/8	6.15	109.9
123	51	12	123.5	2600	2104.8	0.1560	6.7817/6	11.19	123.5
124	75	13	196.2	5454	3257.4	0.3154	11.5556/10	11.56	167.4
125	29	5	16.14	118	94.3	3.9660	2.7115/3	9.97	125.1

Appendix C: Results of the Test for the J Distribution in Taxonomic Data

For each possible combination of taxonomic levels, one of the following tables shows all the Kolmogorov-Smirnov (K-S) scores achieved by each group of organisms. The groups used reflect what was currently available from Internet and library resources. No such source was excluded if it contained a complete—or at least up-to-date—taxonomic synopsis for one high-level group or another.

The first column specifies the group, while the second specifies the range or coverage of the data source for that group. Those field manuals accepted for the study are invariably for North America (NA), but the converse is not true. (See the bibliography.) The third column contains the K-S test results under the heading "score," while the next three columns display critical values of the K-S distribution at the levels of 5%, 10%, and 20%. A score that exceeds any of these values fails the K-S test at that level, and the corresponding critical value is captured in italics to indicate an acceptance failure. The numbers of such failures constitute the results of the test in an overall sense. The column header "# Taxa" indicates the number of lower-level taxa involved in the distribution among higher level ones. These numbers amount to a sample size and are used to compute a normalized score (second last column), as described in Section 4.4.2. Finally, the contribution to overall variance from each individual test is included in the last column. The average normalized score was 0.166.

Species within Genera

Taxonomic Category	Scale	Score	5%	10%	20%	# Taxa	Norm	Var.
Gymnospermatophyta	Global	0.096	0.250	0.232	0.200	27	0.100	0.004
Pteridophyta	NA	0.088	0.123	0.110	0.097	122	0.194	0.001
Angiospermatophyta	NA	0.062	0.070	0.063	*0.055*	3735	0.239	0.005
Bacteria	Global	0.061	0.062	*0.056*	*0.049*	483	0.268	0.010
Aves	Global	0.058	0.094	0.085	0.074	2077	0.413	0.061
Arachnida	NA	0.080	0.091	0.082	*0.071*	224	0.272	0.011
Collembola	Global	0.056	0.060	*0.054*	*0.047*	519	0.273	0.011
Testudines	Global	0.065	0.148	0.133	0.117	84	0.119	0.002
Herpetofauna	NA	0.049	0.122	0.109	0.096	125	0.110	0.003
Mammalia	NA	0.081	0.084	*0.075*	*0.066*	265	0.287	0.015

Species within Families

Taxonomic Category	Scale	Score	5%	10%	20%	No. spp.	Norm	Var.
Gymnospermatophyta	NA	0.191	0.432	0.388	0.339	25	0.270	0.011
Pteridophyta	NA	0.161	0.270	0.240	0.214	122	0.194	0.001
Angiospermatophyta	NA	0.092	*0.089*	*0.080*	*0.070*	233	0.281	0.013
Aves	Global	0.089	0.114	0.102	0.089	143	0.213	0.002
Arachnida	Global	0.094	0.207	0.186	0.163	43	0.123	0.002
Collembola	Global	0.267	*0.240*	*0.220*	*0.193*	30	0.292	0.016
Testudines	Global	0.199	0.361	0.325	0.284	13	0.143	0.001
Herpetofauna	NA	0.084	0.224	0.201	0.176	37	0.102	0.004
Mammalia	NA	0.087	0.182	0.163	0.143	56	0.130	0.001

Species within Orders

Taxonomic Category	Scale	Score	5%	10%	20%	No. spp.	Norm	Var.
Aves	Global	0.209	0.340	0.250	0.220	23	0.200	0.001
Pisces	Global	0.075	0.192	0.173	0.151	50	0.106	0.004
Mammalia	NA	0.169	0.361	0.338	0.295	13	0.122	0.002

Species within Phyla

Taxonomic Category	Scale	Score	5%	10%	20%	No. spp.	Norm	Var.
Bacteria	NA	0.093	0.280	0.214	0.185	33	0.322	0.024

Genera within Families

Taxonomic Category	Scale	Score	5%	10%	20%	No. spp.	Norm	Var.
Gymnospermatophyta	NA	0.122	0.432	0.388	0.339	9	0.073	0.009
Pteridophyta	NA	0.128	0.270	0.240	0.210	25	0.128	0.001
Angiospermatophyta	NA	0.098	*0.088*	*0.079*	*0.070*	237	0.302	0.018
Aves	Global	0.061	0.114	0.102	0.089	143	0.146	0.000
Foraminifera	Global	0.067	0.109	0.097	0.085	157	0.174	0.000
Pisces	Global	0.054	0.065	0.058	*0.051*	439	0.226	0.004
Arachnida	NA	0.029	0.205	0.184	0.161	44	0.038	0.016
Collembola	Global	0.186	0.290	0.220	0.190	30	0.204	0.001
Pogonophora	Global	0.182	0.349	0.314	0.274	14	0.136	0.001
Testudines	Global	0.108	0.361	0.325	0.284	13	0.078	0.008
Herpetofauna	NA	0.089	0.224	0.201	0.176	37	0.108	0.003
Mammalia	NA	0.065	0.179	0.160	0.092	58	0.099	0.004
Ciliophora	Global	0.056	0.082	0.074	0.065	273	0.254	0.008

Genera within Orders

Taxonomic Category	Scale	Score	5%	10%	20%	No. spp.	Norm	Var.
Mammalia	NA	0.218	0.360	0.325	0.284	13	0.157	0.000
Pisces	Global	0.069	0.192	0.173	0.151	50	0.098	0.005
Aves	Global	0.125	0.330	0.250	0.220	23	0.120	0.002
Ciliophora	Global	0.077	0.180	0.162	0.142	57	0.116	0.003

Genera within Phyla

Taxonomic Category	Scale	Score	5%	10%	20%	No. spp.	Norm	Var.
Bacteria	Global	0.072	0.290	0.216	0.185	32	0.081	0.007

Families within Orders

Taxonomic Category	Scale	Score	5%	10%	20%	No. spp.	Norm	Var.
Fungi	Global	0.128	0.142	*0.127*	*0.112*	92	0.246	0.006
Plantae	Global	0.068	0.086	0.077	*0.067*	252	0.216	0.003
Aves	Global	0.134	0.300	0.250	0.220	23	0.129	0.001
Pisces	Global	0.125	0.192	0.173	0.151	50	0.177	0.000
Insecta	NA	0.071	0.270	0.224	0.200	26	0.072	0.009
Ciliophora	Global	0.063	0.180	0.162	0.142	57	0.095	0.005

Families within Classes

Taxonomic Category	Scale	Score	5%	10%	20%	No. spp.	Norm	Var.
Fungi	Global	0.125	0.270	0.245	0.220	24	0.122	0.002
Plantae	Global	0.077	0.280	0.214	0.185	33	0.088	0.006

Orders within Classes

Taxonomic Category	Scale	Score	5%	10%	20%	No. spp.	Norm	Var.
Fungi	Global	0.071	0.270	0.250	0.220	23	0.068	0.010
Animalia	Global	0.067	0.142	0.127	0.112	92	0.129	0.001
Plantae	Global	0.108	0.260	0.232	0.202	27	0.112	0.003

Orders within Phyla and Classes within Phyla (resp.)

Taxonomic Category	Scale	Score	5%	10%	20%	No. spp.	Norm	Var.
Animalia	Global	0.127	0.250	0.232	0.200	27	0.132	0.001

Classes within Phyla and Phyla within Life (resp.)

Taxonomic Category	Scale	Score	5%	10%	20%	No. spp.	Norm	Var.
Animalia	Global	0.071	0.250	0.228	0.200	28	0.075	0.008
Life	Global	0.103	0.143	0.129	0.113	90	0.195	0.001

Bibliography

This bibliography is divided into three sections. The print references section (immediately below) contains all literature pertaining to the expository text of the monograph. The next section, "Web References," contains all references to websites and pertains mainly to Chapter 9 on fossil J-curves. A third section consists of "Meta-study References" used in the meta-study, as described mainly in Chapter 8.

TEXT (PRINT) REFERENCES

Akaike, H. 1974. A new look at statistical model identification. *IEEE Trans. on Automatic Control.* **19** (6): 716–723.

Alpen, E. L. 1998. *Radiation Biophysics.* Academic Press, San Diego, CA.

Balakrishnan, N. 1991. *Handbook of the Logistic Distribution.* Marcel Dekker, Inc., New York.

Barndorff-Nielsen, O. 1978. *Information and Exponential Families in Statistical Theory.* Wiley Series in Probability and Mathematical Statistics. John Wiley & Sons, Ltd., Chichester, England.

Barun, K. and Gupta, S. 1999. *Modern Foraminifera.* Kluwer Academic, Dordrecht, the Netherlands.

Berger, W. H. and Parker, F. L. 1970. Diversity of planktonic foraminifera in deep sea sediments. *Science.* **168**: 1345–1347.

Borrer, D. J. and White, R. E. 1970. *A Field Guide to the Insects: North America North of Mexico.* The Peterson Field Guide Series. Houghton Mifflin, Boston.

Botkin, D. B. 1990. *Discordant Harmonies.* Oxford University Press, New York.

Boulinier, T., Nichols, J. D., Sauer, J. R., Hines, J. E., and Pollock, K. H. 1998. Estimating species richness: The importance of heterogeneity in species detectability. *Ecology.* **79** (3): 1018–1028.

Braun, E. L. 1972. *Deciduous Forests of Eastern North America.* Hefner Publishing Company, New York.

Burlando, B. 1990. The fractal dimension of taxonomic systems. *J. Theor. Biol.* **146**: 99–114.

Burlando, B. 1993. The fractal geometry of evolution. *J. Theor. Biol.* **163**: 161–172.

Burnham, K. P. and Anderson, D. R. 2002. *Model Selection and Multimodel Inference: A Practical Information-Theoretic Approach* (2nd ed.). Springer-Verlag, London.

Burnham, K. P. and Overton, W. S. 1979. Robust estimation of population size when capture probabilities vary among animals. *Ecology.* **60**: 927–936.

Cairns Jr., J. 1969. Factors affecting the number of species in freshwater protozoan communities. In *The Structure and Function of Freshwater Microbial Communities.* Research Division Monographs (J. Cairns Jr., ed.). pp. 219–247. Virginia Polytechnic Institute and State University, Blacksburg, VA.

Cappuccino, N. and Price, P. W., eds. 1995. *Population Dynamics: New Approaches and Synthesis.* Academic Press, San Diego, CA.

Caston, B. J. 1972. *How to Know the Spiders* (3rd ed.). Pictured Key Nature Series. William C. Brown-McGraw Hill, Boston.

Chaitin, G. J. 2001. *Exploring Randomness.* Springer-Verlag, London, England.

Chao, A., Hsieh, T. C., Chazdon, R. L., Colwell, R. K., and Gotelli, N. J. 2015. Unveiling the species-rank abundance distribution by generalizing the Good-Turing sample coverage theory. *Ecology.* **96** (5): 1189–1201.

Charnov, E. L. and Finerty, J. P. 1980. Vole population cycles: A case for kin selection? *Oecologia.* **45**: 1. doi:10.1007/BF00346698.

Chesson, P. 1978. Predator-Prey Theory and Variability. *Annual Review of Ecology and Systematics.* **9**: 323–347.

Chu, J. and Adami, C. 1999. A simple explanation for taxon abundance patterns. *Proc. Natl. Acad. Sci.* **96** (1999): 15017–15019.

Clifford, H. T. and Stephenson, W. 1975. *An Introduction to Numerical Classification.* Academic Press, London, England.

Coddington, J. A., Agnarsson, I., Miller, J. A., Kuntner, M., and Hormiga, G. 2009. Understanding bias: The null hypothesis for singleton species in tropical arthropod surveys. *J. Anim. Ecol.* **78**: 573–584.

Conant, R. and Collins, J. T. 1991. *A Field Guide to Reptiles and Amphibians: Eastern and Central North America* (3rd ed.). Houghton Mifflin, Boston.

Connolly, S. R., Hughes, T. P., Bellwood, D. R., and Karlson, R. H. 2005. Community structure of corals and reef fishes at multiple scales. *Science.* **26** (309): 1363–1365.

DeAngelis, D. L. and Gross, L. J., eds. 1992. *Individual-Based Models and Approaches in Ecology.* Chapman & Hall, New York.

Den Boer, P. J. 1968. Spreading of risk and the stabilization of animal numbers. *Acta Biotheor.* **18**: 165–194.

Den Boer, P. J. 1990. Seeing the trees for the wood: Random walks or bounded fluctuations of population size? *Oecologia.* **86** (4): 484–491.

Deserud, O. and Engen, S. 2000. A general and dynamic species abundance model, embracing the lognormal and gamma models, *Am. Nat.* **155**: 497–511.

Dewdney, A. K. 1996. *Micro-ecology in a Small Stream.* Occasional Report, Environmental Science Program. University of Western Ontario, London, Canada.

Dewdney, A. K. 1998a. A dynamical model of abundances in natural communities. *Coenoses.* **12** (2–3): 67–76.

Dewdney, A. K. 1998b. A general theory of the sampling process with applications to the "veil line." *Theor. Popul. Biol.* **54**: 294–302.

Dewdney, A. K. 2000. A dynamical model of communities and a new species-abundance distribution. *Biol. Bull.* **198** (1): 152–163.

Dewdney, A. K. 2001. The forest and the trees: Romancing the J-curve. *Math. Intell.* **23** (3): 27–34.

Dewdney, A. K. 2003. The stochastic community and the logistic-J distribution. *Acta Oecologica.* **24**: 221–229.

Dewdney, A. K. 2010. The structure of benthic microbial communities in the Old Ausable River Channel. In *Impacts of Water Diversion on Biotic Communities of a River in a Dune Watershed* (M. A. Maun and R. A. Schincariol, eds.). pp. 61–86. NRC Press, Ottawa, ON, Canada.

Doak, D. D. et al. 2008. Understanding and predicting ecological dynamics: Are major surprises inevitable? *Ecology.* **89** (4): 952–961.

Dornelas, M., Connolly, S. R., and Hughes, G. P. 2006. Coral reef diversity refutes the neutral theory of biodiversity. *Lett. Nat.* **440**: 80–82.

Drakare, S., Lennon, J. L., and Hillebrand, H. 2006. The imprint of the geographical, evolutionary and ecological context on species–area relationships. *Ecol. Lett.* **9** (2): 215–227.

Eldredge, N. and Gould, S. J. 1972. Punctuated equilibria: An alternative to phyletic gradualism. In *Models in Paleobiology* (Schopf, T. J. M. ed.). pp. 82–115. Freeman, Cooper and Co., San Francisco.

Emshoff, J. R. and Sisson, R. L. 1970. *Design and Use of Computer Simulation Models*. The MacMillan Company, New York.

Erwin, T. L. 1988. The tropical forest canopy: The heart of biotic diversity. In *Biodiversity* (E. O. Wilson, ed.). pp. 123–129. National Academy Press, Washington, DC.

Erwin, T. L. 1997. Biodiversity at its utmost: Tropical Forest Beetles. In *Biodiversity II* (Reaka-Kudla, M. L., Wilson, D. E., and Wilson, E. O., eds.). pp. 41–68. Joseph Henry Press, Washington, DC.

Feller, W. 1968. *An Introduction to Probability Theory and Its Applications* (Vol. 1). Wiley, New York.

Fisher, R. A., Corbet, A. S., and Williams, C. B. 1943. The relation between the number of species and the number of individuals in a random sample of an animal population. *J. Anim. Ecol.* **12**: 42–58.

Folland, G. B. 1999. *Real Analysis: Modern Techniques and Their Applications* (2nd ed.). Wiley, New York.

Galsworthy, P. 2004. Personal communication to A. K. Dewdney.

Gaston, K. J. 1996. *Biodiversity: A Biology of Numbers and Difference*. Blackwell Science, Oxford, UK.

Gaston, K. J. 2005. The lognormal distribution is not an appropriate null hypothesis for the species-abundance distribution. *J. Anim. Ecol.* **74** (3): 409–433.

Gause, G. F. 1934. *The Struggle for Existence*. Williams & Wilkins Co., Baltimore.

Genet, R. M., Hayes, D. S., Hall, D. S., and Genet, D. R. 1988. *Supernova 1097A*. Fairborn Press, Mesa, AZ.

Gleick, J. 1987. *Chaos: Making A New Science*. Viking Penguin, New York.

Good, I. J. 1953.The population frequencies of species and the estimation of population parameters. *Biometrika*. **40** (3–4): 237–264.

Goodman, L. A. 1949. On the estimation of classes in a population. *Ann. Math. Stat.* **20** (4): 572–579.

Gould, S. J. 2009. *Punctuated Equilibrium*. Harvard University Press, Cambridge, MA.

Gray, J. S. 1988. Species abundance patterns. In *Organization of Communities: Past and Present* (Gee, J. H. R. and Giller, P. S. eds.). Blackwell, Oxford.

Green, J. L. and Plotkin, J. B. 2007. A statistical theory for sampling species abundances. *Ecology Letters*. **10**: 1037–1045.

Griffen, B. D. and Drake, J. M. 2008. A review of extinction in experimental populations. *J. Anim. Ecol.* **77** (6): 1274–1287.

Gross, J. and Yellen, J. 2004. *Handbook of Graph Theory*. CRC Press, Boca Raton, FL.

Hall, R. E. 1981. *The Mammals of North America* (Vol. 1 and 2). Wiley, New York.

Hays, W. L. and Winkler, R. L. 1971. *Statistics: Probability, Inference, and Decision*. Holt, Rinehart, and Winston, New York.

Herbert, N. 1985. *Quantum Reality: Beyond the New Physics*. Doubleday Anchor Books, New York.

Hilborn, R. and Mangel, M. 1997. *The Ecological Detective: Confronting Models with Data*. Princeton U. Press, Princeton, NJ.

Hiyama, A. et al. 2012. The biological impacts of the Fukushima nuclear accident on the Pale Grass Blue Butterfly. *Sci. Rep.* **2**, No. 570.

Hobson, E. 1950. *Theory of Functions*. Harren Press, Washington, DC.

Holt, J. G., Krieg, N. R., Sneath, P. H. A., Staley, J. T., and Williams, S. T. 1994. *Bergey's Manual of Systematic Bacteriology* (9th ed.). Williams & Wilkins, Baltimore.

Hubbell, S. P. 2001. *The Unified Neutral Theory of Biodiversity and Biogeography*. Princeton University Press, Princeton, NJ.

Hughes, R. G. 1984. A model of the structure and dynamics of benthic marine invertebrate communities. *Marine Ecol. Prog. Ser.* **15**: 1–11.

Hughes, R. G. 1986. Theories and models of species abundance. *Am. Nat.* **128** (6): 879–899.

Hurlbert, S. H. 1971. The non-concept of species diversity: A critique and alternative parameters. *Ecology.* **52**: 577–586.

Huston, M. A. 1994. *Biological Diversity.* Cambridge University Press, Cambridge.

IAEA. 2010. *Radiation Biology: A Handbook for Teachers and Students.* International Atomic Energy Agency, Vienna, Austria. (See also web references.)

Ivanov, A. B. 1963. *Pognophora.* Academic Press, London, England.

Iverson, J. B. 1992. *A Revised Checklist with Distribution Maps of the Turtles of the World.* Department of Biology, Earlham College, Richmond, IN.

Jacobs C. G. C. et al. 2014. Egg survival is reduced by grave-soil microbes in the carrion beetle, Nicrophorus vespilloides. *BMC Evolutionary Biology.* **14**: 208.

Jansen, M. 2005. The Lepidoptera fauna of three brackish salt marshes including two species new for the Belgian fauna. *Phegea.* **33** (2).

Johnson, N. L. and Kotz, S. 1970. *Distributions in Statistics: Continuous Univariate Distributions—1.* John Wiley & Sons, New York.

Joseph, F. 1984. *Fishes of the World* (2nd ed.). Wiley, New York.

Judson, O. P. 1994. The rise of the individual-based model in ecology. *Trends Ecol. Evol.* **9** (1): 9–14.

Kartesz, J. T. 1994. *A Synonymized Checklist of the Vascular Flora of the United States, Canada, and Greenland* (2nd ed., Vol. 1). Timber Press, Portland, OR.

Koetsier, P., Dey, P. D., Mladenka, G., and Check, J. W. 1990. Rejecting equilibrium theory: A cautionary note. *Bull. Ecol. Soc. Am.* **71**: 229–230.

Lande, R., Engen, S., and Saether, B.-E. 2003. *Stochastic Population Dynamics in Ecology and Conservation.* Oxford University Press, New York.

Law, A. M. and Kelton, W. D. 1999. *Simulation Modeling and Analysis.* McGraw-Hill Higher Education, New York.

Layberry, R. A., Hall, P. W., and Lafontaine, J. D. 1998. *The Butterflies of Canada.* University of Toronto Press, Toronto, ON, Canada.

Leigh, E. R. 1968. The ecological role of Volterra's equations. In *Some Mathematical Problems in Biology. Lectures on Mathematics in the Life Sciences 1* (American Mathematical Society, ed.). pp. 1–14. American Mathematical Society, Providence, RI.

Levins, R. 1969. The effects of random variations of different types on population growth. *Proceedings of the National Academy of Sciences.* April 1. **62** (4).

Limpert, E., Stahel, W., and Abbot, M. 2001. Log-normal distributions across the sciences: Keys and clues. *Bioscience.* **51**: 341–352.

Loeblich, A. R. and Tappan, H. 1988. *Foraminiferal Genera and Their Classification.* Van Nostrand Reinhold, New York.

Lorenz, E. N. 1963. Deterministic nonperiodic flow. *J. Atmos. Sci.* **20**: 130–141.

MacArthur, R. H. and Wilson, E. O. 1967. *The Theory of Island Biogeography.* Princeton U. Press, Princeton, NJ.

MacArthur, R. H. 1957. On the relative abundance of bird species. *Proc. Natl. Acad. Sci. USA.* **43**: 293–295.

MacArthur, R. H. 1960. On the relative abundance of species. *Am. Nat.* **94**: 25–36.

MacLulich, D. A. 1937. Fluctuations in the numbers of the Varying Hare. *University of Toronto Studies, Biological Series.* No. 43. University of Toronto Press, Toronto, ON, Canada.

Madsen, H. 1982. *Snail Ecology. I: Methodology.* Danish Bilharziasis Laboratory, Charlottenlund, Denmark.

Magurran, A. E. 1988. *Ecological Diversity and Its Measurement.* Croom Helm, London, England.

Maisel, H. and Gnugnoli, G. 1972. *Simulation of Discrete Stochastic Systems.* Science Research Associates, Chicago.

Margulis, L., Corliss, J. O., and Melkonian, M., eds. 1990. *Handbook of Protoctista*. Jones and Bartlett, Boston.

Margulis, L. and Schwartz, K. V. 1982. *Five Kingdoms: An Illustrated Guide to the Phyla of Life on Earth*. Freeman, New York.

Marsh, G. P. 1865. *Man and Nature*. Charles Scribner, New York.

Massey, F. J., Jr. 1951. The Kolmogorov-Smirnov test for goodness of fit. *J. Am. Stat. Assoc.* **465**: 68–78.

Mayr, E. 1970. *Populations, Species, and Evolution*. Harvard University Press, Cambridge, MA.

May, R. M. 1974. Biological populations with nonoverlapping generations: Stable points, stable cycles, and chaos. *Science*. **186**: 645–647.

May, R. M. 1975. Patterns of species abundance and diversity. In *Ecology and Evolution of Communities* (Cody, M. L. and Diamond, J. M., eds.). pp. 81–120. Harvard U. Press, Cambridge, MA.

May, R. M. 1976. Simple mathematical models with very complicated dynamics. *Nature*. **261**: 459.

May, R. M. 1992. How many species inhabit the Earth? *Sci. Am.* **October**: 42–48.

May, R. M. 1998. The voles of Hokkaido. *Nature*. **396**: 409–410.

May, R. M. and Stumpf, M. P. H. 2000. Species–area relationships in tropical forests. *Science*. **290**: 2084–2086.

McGill, B. J. et al. 2007. Species abundance distributions: Moving beyond single theories to integration within an ecological framework. *Ecol. Lett.* **10**: 995–1015.

Morris, W. et al. 1999. *A Practical Handbook for Population Viability Analysis*. The Nature Conservancy, Arlington County, VA.

Morters, P., and Peres, Y. 2010. *Brownian Motion*. Cambridge Series in Statistical and Probabilistic Mathematics, Cambridge University Press, Cambridge, UK.

Muller, H. J. 1927. Artificial transmutation of the gene. *Science*. **66** (169): 84–87.

Murdoch, W. W. 1994. Population regulation in theory and practice. *Ecology*. **75** (2): 271–287.

Nelson, J. S., Grande, T. C., and Wilson, M. V. H. 2016. *Fishes of the World* (5th ed.). Wiley, New York.

Neuts, M. F. 1973. *Probability*. Allyn and Bacon, Boston.

Odum, E. 1953. *Fundamentals of Ecology*. W. B. Saunders, Toronto, ON, Canada.

Page, L. M. and Burr, B. M. 1991. *A Field Guide to the Freshwater Fishes*. Houghton Mifflin Company, Boston.

Pearson, A. V. and Hartley, H. O. 1972. *Biometrica Tables for Statisticians* (2nd ed.). Cambridge University Press, Cambridge, UK.

Pearson, K. 1900. On a criterion that a given system of deviations from the probable in the case of a correlated system of variables is such that it can be reasonably supposed to have arisen in random sampling. *Phil. Mag. 5th Ser.* **50**: 157–175.

Peters, R. H. 1995. *A Critique for Ecology*. Cambridge University Press, Cambridge, UK.

Pielou, E. C. 1969. *An Introduction to Mathematical Ecology*. Wiley, New York.

Pielou, E. C. 1977. *Mathematical Ecology*. Wiley, New York.

Press, W. H., Flannery, B. P., Teukolsky, S. A., and Vetterling, W. T. 1986. *Numerical Recipes: The Art of Scientific Computing*. Cambridge University Press, Cambridge, UK.

Preston, F. W. 1948. The commonness, and rarity, of species. *Ecology*. **29**: 254–283.

Preston, F. W. 1962. The canonical distribution of commonness and rarity: Part I. *Ecology*. **43**: 185–215 and 431–432.

Rae, A. 1986. *Quantum Physics: Illusion or Reality?* Cambridge University Press, Cambridge, UK.

Raup, D. M. and Stanley, S. M. 1978. *Principles of Paleontology*. W. H. Freeman, New York.

Rice, W. R. and Hostert, E. E. 1993. Laboratory experiments on speciation: What have we learned in 40 years? *Evolution*. **47** (8): 1637–1653.

Ricklefs, R. E. 1990. *Ecology*. W. H. Freeman, New York.

Ridley, M. 1996. *Evolution* (2nd ed.). Blackwell Science, Inc., Cambridge, MA.

Robbins, C. S., Brunn, B., and Zim, H. S. 1983. *Birds of North America*. Golden Press, New York.

Robbins, C. S., Bystrak, D., and Geissler, P. H. 1986. *The Breeding Bird Survey: Its First Fifteen Years, 1965–1979*. No. FWS-PUB-157. Patuxent Wildlife Research Centre, Laurel, MD.

Rosenzweig, M. L. 1995. *Species Diversity in Space and Time*. Cambridge University Press, Cambridge, UK.

Rothschild, L. J. 1999. The influence of UV radiation on Protistan evolution. *Jnl. Eukaryotic Microbiology*. **46** (5).

Shannon, C. E. 1949. *The Mathematical Theory of Communication*. University of Illinois, Urbana, IL.

Sibley, C. G. and Moore, B. L. Jr. 1990. *Distribution and Taxonomy of Birds of The World*. Yale, New Haven, CN.

Simpson, E. H. 1949. Measurement of diversity. *Nature*. **163**: 688.

Smith, E. P. and van Belle, G. 1984. Nonparametric estimation of species richness. *Biometrica*. **40**: 119–129.

Smith, H. S. 1935. The role of biotic factors in the determination of population densities. *J. Econ. Entomol*. **28**: 873–889.

Stenseth, N. C. and Saitoh, T., eds. 1998. *Res.Pop. Ecol*. **40**: 1–158.

Stephenson, R. and Metcalfe, N. H. 2013. Drosophila melanogaster: A fly through its history and current use. *J. R. Coll. Phys. Edinburgh*. **43** (1): 70–75.

Strong, D. R. 1986. Density-vague population change. *Trends Ecol. Evol*. **1** (2): 39–42.

Sugihara, G. 1980. Minimal community structure: An explanation of species abundance patterns. *Am. Nat*. **116**: 770–787.

Svensmark, H. 2012. Evidence of nearby supernovae affecting life on Earth. *Mon. Not. R. Astronom. Soc*. **423** (2): 1234–1253.

Symul, T., Assad, S. M., and Lam, P. K. 2011. Real time demonstration of high bitrate quantum random number generation with coherent laser light. *Appl. Phys. Lett*. **98**: 231103 (on web).

Trivedi, K. S. 2001. *Probability and Statistics with Reliability, Queueing, and Computer Science Applications* (2nd ed.). John Wiley, New York.

Tsakas, S. C. and David, J. R. 1986. Speciation burst hypothesis: An explanation for the variation in rates of phenotypic evolution. *Genet. Sel. Evol*. **18** (3): 351–358.

Ugland, K. I., Gray, J. S., and Ellingsen, K. E. 2003. The species-accumulation curve and estimation of species richness. *J. Anim. Ecol*. **72** (5): 888–897.

Van Valen, L. M. 1973. A new evolutionary law. *Evol. Theory*. **1**: 1–30.

Verhulst, P. F. 1838. Notice sur la loi que la population poursuit dans son accroissement. *Correspondance mathématique et physique*. **10**: 113–121.

Wallace, B. 1958. The average effect of radiation-induced mutations on viability in Drosophila melanogaster. *Evolution*. **12** (4): 532–556.

Wallace, B. 1981. *Basic Population Genetics*. Columbia University Press, New York.

Watt, K. E. F. 1962. Use of mathematics in population ecology. *Ann. Rev. Entomol*. **7**: 243–260.

Waters, T. F. 1999. Long-term trout production dynamics in Valley Creek, Minnesota. *Trans. Am. Fish. Soc*. **128**: 1151–1162.

Whitfield, J. 2003. Genome pioneer sets sights on Sargasso Sea. *Nature*. 30 April (online).

Whittaker, J. O. Jr. 1996. *National Audubon Society Field Guide to the North American Mammals*. Knopf, New York.

Whittaker, R. H. 1977. Evolution of species diversity in land communities. In *Evolutionary Biology* (Vol. 10) (Hecht, M. K., Steere, W. C., and Wallace, B., eds.). pp. 1–67. Plenum Press, New York.

Williams, C. B. 1943. Area and the number of species. *Nature.* **152**: 264–267.

Williams, C. B. 1964. *Patterns in the Balance of Nature.* Academic Press, London, England.

Williamson, D. I. 2009. Caterpillars evolved from Onychophorans by hybridogenesis. *Proc. Natl. Acad. Sci.* **106** (47): 19901–19905.

Willis, J. C. and Yule, G. U. 1922. Some statistics of evolution and geographic distribution in plants and animals and their significance. *Nature.* **109**: 177.

Wilson, E. O., Eisner, T., Briggs, W. R., Dickerson, R. E., Metzenberg, R. L., O'Brien, R. D., Susman, M., and Boggs, W. E. 1973. *Life on Earth.* Sinauer Assoc., Stamford, CT.

Wolda, H. 1995. The demise of the population regulation controversy. *Researches on Population Ecology.* **37**: 91–93.

Yule, G. U. 1924. A mathematical theory of evolution, based on the conclusions of Dr. J. C. Willis, F. R. S. *Phil. Trans. Roy. Soc. London.* Ser. B. **213**: 21–87.

WEB REFERENCES

Bellinger, P. F., Christiansen, K. A., and Janssens, F. 1996–2003. Checklist of the Collembola. http://webhost.ua.ac.be/collembola/biogeo/index.php and http://www.collembola.org

Chao, A. 2005. Species Richness Estimation. http://viceroy.eeb.uconn.edu/EstimateSPages/EstSUsersGuide/References/Chao2005.pdf

Chu, J. and Adami, C. 1999. A simple explanation for taxon abundance distributions. *Proc. Natl. Acad. Sci.* **96** (26): 15017. www.pnas.org/cgi/full/96/26/15017

Cutler, E. B. 1994. Cutler's Sipuncula. Harvard University website. http://www.mcz.harvard.edu/Departments/InvertZoo/as.fldr/cutler/sipclass.htm

Dewdney, A. K. 2009. A bibliography for the 125-biosurvey metstudy. http://www.csd.uwo.ca/~akd/environment/biosurvs2.html

Dewdney, A. K. 2011. Biodiversity Research. http://www.csd.uwo.ca/faculty/~akd/. Click on "test data."

Hatfield, B. and Tinker, T. 2013. Spring 2013 California Sea Otter Census Results, United States Geological Survey, September 12. http://www.werc.usgs.gov/seaottercount

IAEA. 2010. *Radiation Biology: A Handbook for Teachers and Students.* International Atomic Energy Agency, Vienna, Austria. http://www-pub.iaea.org/MTCD/publications/PDF/TCS-42_web.pdf

ITIS. 2002a. *Integrated Taxonomic Information System, Hierarchical Report: Fungi.* http://www.itis.usda.gov/itis_phy.html, accessed May 3, 2004. Used: http://www.itis.usda.gov/cgi_bin/itis_phylo_report.cgi

ITIS. 2002b. *Integrated Taxonomic Information System, Hierarchical Report: Plantae.* http://www.itis.usda.gov/cgi_bin/itis_phylo_report.cgi

ITIS. 2002c. *Integrated Taxonomic Information System, Hierarchical Report: Animalia.* http://www.itis.usda.gov/cgi_bin/itis_phylo_report.cgi

Lynn, D. 2007. The Ciliate Resource Archive. http://www.uoguelph.ca/~ciliates/classification/genera.html

Massachusetts Institute of Technology. 2006. Open Course Ware: Mathematics 443 Lecture Notes. http://ocw.mit.edu/courses/mathematics/18-443-statistics-for-applications-fall-2006/lecture-notes/lecture11.pdf

Rafferty, J. P. Little ice age. n.d. *Encyclopedia Britannica.* http://www.britannica.com/science/Little-Ice-Age. Accessed 2016.

Regan, M. 2011. Sampling distribution simulation. University of Aukland. http://www.socr.ucla.edu/Applets.dir/SamplingDistributionApplet.html

Shi, Z. 2010. Random walks & trees. Université Paris. http://www.proba.jussieu.fr/pageperso/zhan/pdffile/guanajuato.pdf

Strong, D. R. 1991. MacArthur Award: William Murdoch. *Bulletin of the Ecological Society of America.* **72** (1): 22–23. www.jstor.org/stable/20167245

Thornton, D. 2013. Bioinformatics, biology and computing. http://dan-thornton.blogspot
    .ca/2013/03/the-exponential-distribution.html
TOL. 2001. The Tree of Life Web Project. http://tolweb.org/tree/
UKSF. 2004. United Kingdom Systematics Forum. http://www.nhm.ac.uk/hosted_sites/uksf
    /uksfd/tall.htm
Wolfram, S. 2016. Galton Board, Wolfram Mathworld. http://mathworld.wolfram.com
    /GaltonBoard.html

META-STUDY REFERENCES

Alaback, P. B. and Herman, F. R. 1988. Long-term response of understory vegetation to stand
    density in Picea-Tsuga forests. *Can. J. For. Res.* **18**: 1522–1530.
Aoki, T. and Tokumasu, S. 1995. Dominance and diversity of the fungal communities on fir
    needles. *J. Microb. Res.* **99** (12): 1439–1449.
Al-Safadi, M. M. 1991. Freshwater macrofauna of stagnant waters in Yemen Arab Republic.
    Hydrobiologica. **210**: 203–208.
Bach, C. E. 1994. Effects of a specialist herbivore (*Altica subplicata*) on *Salix cordata* and
    sand dune succession. *Ecol. Monogr.* **64** (4): 423–445.
Barlocher, F. K. B. 1974. Dynamics of the fungal population on leaves in a stream. *J. Ecol.*
    62: 761–791.
Baz, A. and Garcia-Boyero, A. 1995. The effects of forest fragmentation on butterfly com-
    munities in central Spain. *J. Biogeogr.* **22**: 120–140.
Benstead, J. P. 1996. Macroinvertebrates and the processing of leaf litter in a tropical stream.
    Biotropica. **28** (3): 367–375.
Bersier, L.-F. and Meyer, D. R. 1994. Bird assemblages in mosaic forests: The relative impor-
    tance of vegetation structure and floristic composition along the successional gradient.
    Acta Qecologia. **15** (5): 561–576.
Bills, G. F. and Polishook, J. D. 1994. Abundance and diversity of microfungi in leaf litter of
    a lowland rain forest in Costa Rica. *Mycologia.* **86** (2): 187–198.
Bohning-Gaese, K. and Bauer, H.-G. 1996. Changes in species abundance, distribution, and
    diversity in a central European bird community. *Conserv. Biol.* **10**: 175–187.
Brosset, A., Charles-Dominique, P., Cockle, A., Cosson, J.-F., and Masson, D. 1996. Bat com-
    munities and deforestation in French Guiana. *Can. J. Zool.* **74**: 1974–1982.
Brothers, T. S. 1993. Fragmentation and edge effects in Central Indiana old-growth forests.
    Nat. Areas J. **13** (4): 268–274.
Brunner, I., Brunner, F., and Laursen, G. A. 1992. Characterization and comparison of mac-
    rofungal communities in an *Alnus tenuifolia* and an *Alnus crispa* forest in Alaska. *Can.
    J. Bot.* **70**: 1247–1258.
Busby, W. H. and Parmelee, J. R. 1996. Historical changes in a herpetofaunal assemblage in
    the Flint Hills of Kansas. *Am. Midl. Nat.* **135**: 81–91.
Butler, L. 1992. The community of macrolepidopterous larvae at Cooper's Rock State Forest,
    West Virginia: A baseline study. *Can. Entomol.* **124**: 1149–1156.
Cairns, J. Jr. and Ruthven, J. A. 1972. A test of the cosmopolitan distribution of freshwater
    protozoans. *Hydrobiologica.* **39** (3): 405–427.
Caley, M. J. 1995. Community dynamics of tropical reef fishes: Local patterns between lati-
    tudes. *Mar. Ecol. Prog. Ser.* **129**: 7–18.
Catling, P. M. and Lefkovitch, L. P. 1989. Associations of vascular epiphytes in a Guatemalan
    cloud forest. *Biotropica.* **21** (1): 35–40.
Chancellor, R. J. 1985. Changes in the weed flora of an arable field cultivated for 20 years.
    J. Appl. Ecol. **22**: 491–501.

Chandler, D. S. and Peck, S. B. 1992. Diversity and seasonality of Leiodid beetles (Coleoptera: Leiodidae) in an old-growth and a 40-year old forest in New Hampshire. *Environ. Entomol.* **21** (6): 1283–1293.

Chester, E. W. and Noel, S. M. 1995. A phytosociological analysis of an old-growth upland wet woods on the Pennyroyal Plain, southcentral Kentucky, USA. *Nat. Area J.* **15** (4): 297–307.

Christiansen, M. B. and Pitter, E. 1997. Species loss in a forest bird community near Lagoa Santa in southeastern Brazil. *Biol. Cons.* **80**: 23–32.

Colwell, M. A. and Dodd, S. L. 1995. Waterbird communities and habitat relationships in coastal pastures of northern California. *Cons. Biol.* **9** (4): 827–834.

Cowx, I. G., Young, W. O., and Hellawell, J. M. 1984. The influence of drought on the fish and invertebrate populations of an upland stream in Wales. *Freshwater Biol.* **14**: 165–177.

Deniseger, J., Austin, A., and Lucey, W. P. 1986. Periphyton communities in a pristine mountain stream above and below heavy metal mining operations. *Freshwater Biol.* **16**: 209–218.

Dewdney, A. K. 1996. *Ecology in a Small Stream.* Monograph. Environmental Science Program, UWO, London, Ontario, Canada.

Edgar, G. J. and Shaw, C. 1995. The production and trophic ecology of shallow-water fish assemblages in southern Australia 1. Species richness, size-structure and production of fishes in Western Port, Victoria. *J. Exp. Mar. Biol. and Ecol.* **194**: 53–81.

Estrada, A., Coates-Estrada, R., and Merritt, D. Jr. 1994. Non flying mammals and landscape changes in the tropical rain forest region of Los Tuxtlas, Mexico. *Ecography.* **17**: 229–241.

Fedynich, A. M., Pence, D. B., Gray, P. N., and Bergan, J. F. 1996. Helminth community structure and pattern in two allopatric populations of a nonmigratory waterfowl species (*Anas fulvigula*). *Can. J. Zool.* **74**: 1253–1259.

Finnamore, B. 1998. Personal communication.

Flanagan, G. J., Wilson, C. G., and Gillett, J. D. 1990. The abundance of native insects on the introduced weed *Mimosa pigra* in Northern Australia. *J. Trop. Ecol.* **6**: 219–230.

Fox, M. D. and Fox, B. J. 1986. The effect of fire on the structure and floristic composition of a woodland understorey. *Can. J. Ecol.* **11**: 77–85.

Frank, T. and Nentwig, W. 1995. Ground dwelling spiders (Araneae) in sown weed strips and adjacent fields. *Acta Oecologia.* **16** (2): 179–193.

Galacatos, K., Stewart, D. J., and Ibarra, M. 1996. Fish community patterns of lagoons and associated tributaries in the Equadorian Amazon. *Copeia.* **4**: 875–894.

Garcia-Raso, J. E. 1990. Study of a Crustacea Decapoda Taxocoenosis of *Poisidonia oceanica* beds from the southeast of Spain. *Mar. Ecol.* **11** (4): 309–326.

Getz, L. L. and Uetz, G. W. 1994. Species diversity of terrestrial snails in the southern Appalachian mountains. *Malacol. Rev.* **27**: 61–74.

Gorchov, D. L., Cornejo, F., Ascorra, C. F., and Jaramillo, M. 1995. Dietary overlap between frugivorous birds and bats in the Peruvian Amazon. *OIKOS.* **74**: 235–250.

Griffiths, R. W. 1991. Environmental quality assessment of the St. Clair River as reflected by the distribution of benthic macroinvertebrates in 1985. *Hydrobiologica.* **219**: 143–164.

Griffiths, R. W., Thornley, S., and Edsall, T. A. 1991. Limnological aspects of the St. Clair River. *Hydrobiologica.* **219**: 97–123.

Gutierrez, D. and Menendez, R. 1995. Distribution and abundance of butterflies in a mountain area in the northern Iberian peninsula. *Ecography.* **18**: 209–216.

Guttierrez, J. R., Meserve, P. L., Contreras, L. C., Vasquez, H., and Jaksic, F. M. 1993. Spatial distribution of soil nutrients and ephemeral plants underneath and outside the canopy of *Porlieria chilensis* shrubs (Zygophyllaceae) in arid coastal Chile. *Oecologia.* **95**: 347–352.

Haila, Y., Jarninen, O., and Vaisanen, R. A. 1980. Habitat distribution and species associations of land bird populations on the Aland Islands, SW Finland. *Ann. Zool. Fennici.* **17**: 87–106.

Harper, F. P. and Harper, F. 1982. Mayfly communities in a Laurentian watershed (insecta; Ephemeroptera). *Can. J. Zool.* **60**: 2828–2840.

Hietz, P. and Hietz-Seifert, U. 1995. Structure and ecology of epiphyte communities of a cloud forest in central Veracruz, Mexico. *J. Veg. Sci.* **6**: 719–728.

Hill, J. K., Hamer, K. C., Lace, L. A., and Banham, W. M. T. 1995. Effects of selective logging on tropical forest butterflies on Buru, Indonesia. *J. Appl. Ecol.* **32**: 754–760.

Holmes, R. T. and Sherry, T. W. 1986. Bird community dynamics in a temperate deciduous forest: Long-term trends at Hubbard Brook. *Ecol. Monogr.* **56** (3): 201–220.

Holmquist, J. G., Powell, G. V. N., and Sogard, S. M. 1989. Decapod and stomatopod assemblages on a system of seagrass-covered mud banks in Florida Bay. *Mar. Biol.* **100**: 473–483.

Hornbach, D. J., Miller, A. C., and Payne, B. S. 1992. Species composition of the mussel assemblages in the upper Mississippi River. *Malacol. Rev.* **25**: 119–128.

Hudon, C. 1994. Biological events during ice breakup in the Great Whale River (Hudson Bay). *Can. J. Fish. Aquat. Sci.* **51**: 2467–2481.

Hulme, P. E. 1996. Herbivores and the performance of grassland plants: A comparison of arthropod, mollusc and rodent herbivory. *J. Ecol.* **84**: 43–51.

Hutto, R. L. 1995. Composition of bird communities following stand-replacement fires in northern Rocky Mountains (U.S.A.) conifer forests. *Conserv. Biol.* **9** (5): 1041–1058.

Janzen, D. H. 1973. Sweep samples of tropical foliage insects: Description of study sites, with data on species abundances and size distributions. *Ecology.* **54** (3): 659–666.

Jefferson, R. G. and Usher, M. B. 1987. The seed bank in soils of disused chalk quarries in the Yorkshire Wolds, England: Implications for conservation management. *Biol. Conserv.* **42**: 287–302.

Jenkins, R. A., Wade, K. R., and Pugh, E. 1984. Macroinvertebrate–habitat relationships in the river Teifi catchment and the significance to conservation. *Freshwater Biol.* **14**: 23–42.

Jennings, S., Brierly, A. S., and Walker, J. W. 1994. The inshore fish assemblies of the Galapagos Archipelago. *Biol. Conserv.* **70**: 49–57.

Johns, A. D. 1991. Responses of Amazonian rain forest birds to habitat modification. *J. Trop. Ecol.* **7**: 417–437.

Kapelle, M., Geuze, T., Leal, M. E., and Cleef, A. M. 1996. Successional age and forest structure in a Costa Rican upper montane Quercu forest. *J. Trop. Ecol.* **12**: 681–698.

Kemp, W. P. 1992. Temporal variation in rangeland grasshopper (Orthoptera: Acrididae) communities in the steppe region of Montana, USA. *Can. Entomol.* **124**: 437–450.

Kingston, T., Akbar, Z., and Kunz, T. H. 2003. Species richness and rarity in an insectivorous bat assemblage from Malaysia. *J. Trop. Ecol.* **18** (01): 67–79.

Koen, J. H. and Crowe, T. M. 1987. Animal–habitat relationships in the Knysna Forest, south Africa: Discrimination between forest types by birds and invertebrates. *Oecologia.* **72**: 414–42.

Krasnov, B., and Shenbrot, G. 1996. Spatial structure of community of darkling beetles (Coleoptera: Tenebrionidae) in the Negev Highlands, Israel. *Ecography.* **19**: 139–152.

Lackey, J. B. 1938. A study of some ecologic factors affecting the distribution of protozoa. *Ecol. Monogr.* **8** (4): 503–527.

Lenanton, R. C. J. and Caputi, N. 1989. The roles of food supply and shelter in the relationship between fishes, in particular Cnidoglanis macrocephalus (Valenciennes) and detached macrophytes in the surf zone of sandy beaches. *J. Exp. Mar. Biol. Ecol.* **128**: 165–176.

Levey, D. J. 1988. Tropical wet forest treefall gaps and distributions of understory birds and plants. *Ecology.* **69** (4): 1076–1089.

Lieberman, D., Lieberman, M., Peralta, R., and Hartshorn, G. S. 1996. Tropical forest structure and composition on a large-scale altitudinal gradient in Costa Rica. *J. Ecol.* **84**: 137–152.

Lowry, J. K. 1975. Soft bottom macrobenthic community of Arthur Harbor in Antarctica. In *Biology of the Antarctic Seas V* (Pawson, D. L., ed.). pp. 10–18. American Geophysical Union, Washington, D.C.

Lundqvist, L. and Brinck-Lindroth, G. 1990. Patterns of coexistence: Ectoparasites on small mammals in northern Fennoscandia. *Holarctic Ecol.* **13**: 39–49.

Maycock, P. F. and Fahselt, D.1992. Vegetation of stressed calcareous screes and slopes in Sverdrup Pass, Ellesmere Island, Canada. *Can. J. Bot.* **70**: 2359–2377.

McCabe, T. L. and Weber, C. N. 1994. The robber flies (Diptera: Asilidae) of the Albany Pinebush. *Great Lakes Entomol.* **27** (3): 157–159.

McIvor, J. G. 1998. Pasture management in semiarid tropical woodlands: Effects on species diversity. *Aus. J. Ecol.* **23**: 349–364.

Medley, K. E. 1992. Patterns of forest diversity along the Tana River, Kenya. *J. Trop. Ecol.* **8**: 353–371.

Menni, R. C., Miquelarena, A. M., Lopez, H. L., Casciotta, J. R., Almiron, A. E., and Protogino, L. C. 1992. Fish fauna and environments of the Pilcomayo-Paraguay basins in Formosa, Argentina. *Hydrobiologica.* **245**: 129–146.

Merrett, N. R., Haedrich, R. L., Gordon, J. D. M., and Stehmann, M. 1991. Deep demersal fish assemblage structure in the Porcupine Seabight (eastern North Atlantic): Results of single warp trawling at lower slope to abyssal soundings. *J. Mar. Biol. Ass.* **71**: 359–373.

Metzger, J. P., Bernacci, L. C., and Goldenberg, R. 1997. Pattern of tree species diversity in riparian forest fragments of different widths (SE Brazil). *Plant Ecol.* **133**: 135–152.

Miller, A. C. and Payne, B. S. 1993. Qualitative versus quantitative sampling to evaluate population and community characteristics at a large-river mussel bed. *Am. Midl. Nat.* **130**: 133–145.

Musil, C. F. and de Witt, D. M. 1990. Post-fire regeneration in a sand plain lowland fynbos community. *S. Afr. J. Bot.* **56**: 167–184.

Neira, F. J., Potter, J. C., and Bradley, J. S. 1992. Seasonal and spatial changes in the larval fish fauna within a large temperate Australian estuary. *Mar. Biol.* **112**: 1–16.

Novak, R. O. and Whittingham, W. F. 1968. Soil and litter microfungi of a maple-elm-ash floodplain community. *Mycologia.* **60**: 776–787.

Niemela, J., Haila, Y., and Halme, E. 1988. Carabid beetles on isolated islands and on the adjacent Aland mainland: Variation in colonization success. *Ann. Zool. Fennici.* **25**: 133–143.

Okali, D. U. U. and Ola-Adams, B. A. 1987. Tree population changes in treated rain forest at Omo Reserve, southwestern Nigeria. *J. Trop. Ecol.* **3**: 291–313.

Paasivirta, L., Lahti, T., and Peratie, T. 1988. Emergence phenology and ecology of aquatic and semi-terrestrial insects on a boreal raised bog in central Finland. *Holarct. Ecol.* **11**: 96–105.

Pajunen, T., Haila, Y., Halme, E., Niemela, J., and Puntilla, P. 1995. Ground-dwelling spiders (Arachnida, Araneae) in fragmented old forests and surrounding managed forests in southern Finland. *Ecography.* **18**: 62–72.

Pearman, P. B. 1997. Correlates of amphibian diversity in an altered landscape of Amazonian Ecuador. *Conserv. Biol.* **11** (5): 1211–1225.

Pucheta, E., Cabido, M., Diaz, S., and Funes, G. 1998. Floristic composition, biomass, and aboveground net plant production in grazed and protected sites in a mountain grassland of central Argentina. *Acta Oecologia.* **19** (2): 97–105.

Rabinowitz, D. and Rapp, J. K. 1985. Colonization and establishment of Missouri prairie plants on artificial soil disturbances. III. Species abundance distributions, survivorship, and rarity. *Am. J. Bot.* **72** (10): 1635–1640.

Rakocinski, C. F., Baltz, D. M., and Fleeger, J. W. 1992. Correspondence between environmental gradients and the community structure of marsh-edge fishes in a Louisiana estuary. *Mar. Ecol. Progr. Ser.* **80**: 135–148.

Rico-Gray, V. and Garcia-Franco, J. G. 1992. Vegetation and soil seed bank of successional stages in tropical lowland deciduous forest. *J. Veg. Sci.* **3**: 617–624.

Rodriguez, M. A. and Lewis, W. M. Jr. 1997. Structure of fish assemblages along environmental gradients in floodplain lakes of the Orinioco River. *Ecol. Monogr.* **67** (1): 109–128.

Roovers, L. M. and Shifley, S. R. 1997. Composition and dynamics of Spitler Woods, an old-growth remnant forest in Illinois (USE). *Nat. Areas J.* **17**: 219–232.

Salvado, H. and Gracia, Ma. d.P. 1991. Response of ciliate populations to changing environmental conditions along a freshwater reservoir. *Archiv. f. Hydrobiologie.* **123** (2): 239–255.

Samson, D. A., Rickart, E. A., and Gonzales, P. C. 1997. Ant diversity and abundance along an elevational gradient in the Philippines. *Biotropica.* **29** (3): 349–363.

Samways, M. J. 1990. Species temporal variability: Epigaeic and assemblages and management for abundance and scarcity. *Oecologia.* **84**: 482–490.

Schmid, P. E. 1992. Community structure of larval Chironomidae (Diptera) in a backwater area of the River Danube. *Freshwater Biol.* **27**: 151–167.

Schmid-Araya, J. M. and Zuniga, L. R. 1992. Zooplankton community structure in two Chilean reservoirs. *Archiv. f. Hydrobiologie.* **123** (3): 305–335.

Serafy, J. E., Lindeman, K. C., Hopkins, T. E., and Ault, J. S. 1997. Effects of freshwater canal discharge on fish assemblages in a subtropical bay: Field and laboratory observations. *Mar. Ecol. Prog. Ser.* **160**: 161–172.

Short, T. M., Black, J. A., and Birge, W. J. 1991. Ecology of a saline stream: Community responses to spatial gradients of environmental conditions. *Hydrobiologia.* **226**: 167–178.

Shorthouse, D. 1998. *The Diversity and Succession of Wandering Spider Communities on INCO Ltd. Reclaimed Tailings Habitats.* PhD thesis, Dept. of Biology, Laurentian University.

Smith, K. L. Jr., Kaufmann, R. S., Edelman, J. L., and Baldwin, R. J. 1992. Abyssopelagic fauna in the central North Pacific: Comparison of acoustic detection and trawl and baited trap collections to 5800 m. *Deep Sea Res.* **39** (3/4): 659–685.

Spanier, E., Pisanty, S., Tom, M., and Almog-Shtayer, G. 1989. The fish assemblage on a coralligenous shallow shelf off the Mediterranean coast of northern Israel. *J. Fish Biol.* **35**: 641–649.

Steenkamp, H. E. and Chown, S. L. 1996. Influence of dense stands of an exotic tree, *Prosopis glandulosa* Benson, on a savannah dung beetle (Coleoptera: Scarabaeinae) assemblage in Southern Africa. *Biol. Conserv.* **78**: 305–311.

Stein, D. L., Tissot, B. N., Hixon, M. A., and Barss, W. 1992. Fish–habitat associations on a deep reef at the edge of the Oregon continental shelf. *Fishery Bull. U.S.* **90**: 540–551.

Stephenson, S. L. 1988. Distribution and ecology of Myxomycetes in temperate forests. I. Patterns of occurrence in the upland forests of southwestern Virginia. *Can. J. Bot.* **66**: 2187–2207.

Sukumar, R., Dattaraja, H. S., Suresh, H. S., Radhakrishnan, J., Vasudeva, R., Nirmala, S., and Joshi, N. V. 1992. Long-term monitoring of vegetation in a tropical deciduous forest in Mudumalai, southern India. *Curr. Sci.* **62**: 608–616.

Taft, J. B., Schwartz, M. W. and Loy, R. P. 1995. Vegetation ecology of flatwoods on the Illinoian till plain. *J. Veg. Sci.* **6**: 647–666.

Tan, T. K., Leong, W. F., and Jones, E. B. G. 1989. Succession of fungi on wood of *Avicennia alba* and *A. lanata* in Singapore. *Can. J. Bot.* **67**: 2686–2691.

Thiollay, J.-M. 1996. The role of traditional agroforests in the conservation of rainforest bird diversity in Sumatra. *Conserv. Biol.* **9** (2): 335–353.

Travnichek, V. H. and Maceina, M. J. 1994. Comparison of flow regulation effects on fish assemblages in shallow and deep water habitats. *J. Freshwater Ecol.* **9** (3): 207–216.

Tuomist, H., and Poulsen, A. D. 1996. Influence of edaphic specialization on pteridophyte distribution in neotropical rain forests. *J. Biogeogr.* **23**: 283–293.

Tzeng, W.-N. and Wang, Y.-T. 1992. Structure, composition and seasonal dynamics of the larval and juvenile fish community in the mangrove estuary of Tanshui River, Taiwan. *Mar. Biol.* **113**: 481–490.

Varga, S. 1965. *Vascular Plant Inventory of the Backus Tract*. The Long Point Region Conservation Authority, Simcoe, Ontario, Canada.

Vuori, K.-M. and Joensuu, I. 1996. Impact of forest drainage on the macroinvertebrates of a small boreal headwater stream: Do buffer zones protect lotic biodiversity? *Biol. Conserv.* **77**: 87–95.

Walters, K. 1991. Influences of abundance, behavior, species composition, and ontogenetic stage on active emergence of meiobenthic copepods in subtropical habitats. *Mar. Biol.* **108**: 207–215.

Wareborn, I. 1992. Changes in the land mollusc fauna and soil chemistry in an inland district in southern Sweden. *Ecography.* **15**: 62–69.

Webb, M. G. 1956. An ecological study of brackish water ciliates. *J. Anim. Ecol.* **25**: 148–175.

Welch, D. 1985. Studies in the grazing of heather moorland in northeast Scotland. *J. Appl. Ecol.* **22**: 461–472.

Whitaker, D. M. and Montevecchi, W. A. 1997. Breeding bird assemblages associated with riparian, interior forest, and nonriparian edge habitats in a balsam fir ecosystem. *Can. J. For. Res.* **27**: 1159–1167.

White, D. H., Kepler, C. B., Hatfield, J. S., Sykes, P. W. Jr., and Seginak, J. T. 1996. Habitat associations of birds in the Georgia Piedmont during winter. *J. Field Ornithol.* **67** (1): 159–166.

Wilson, E. O. 1987. The arboreal ant fauna of Peruvian Amazon forest: A first assessment. *Biotropica.* **19** (3): 245–251.

Winemiller, K. O. and Leslie, M. A. 1992. Fish assemblages across a complex, tropical freshwater/marine ecotone. *Env. Biol. Fish.* **34**: 29–50.

Wright, J. M. 1989. Diel variation and seasonal consistency in the fish assemblage of the nonestuarine Sulaibikhat Bay, Kuwait. *Mar. Biol.* **102**: 135–142.

Yoklavich, M. M., Cailliet, G. M., Barry, J. P., Ambrose, D. A., and Antrim, B. S. 1991. Temporal and spatial patterns in abundance and density of fish assemblages in Elkhorn Slough, California. *Estuaries.* **14**: 465–480.

Endpaper: Global Map of Biosurvey Sites Used in the Meta-Study

Each dot in the world map represents the sampling site for one of the 125 biosurveys used in the meta-study to establish the universality of the J distribution in living communities. Since the selection of studies to incorporate into the meta-study was essentially random, the map gives a rough idea of the global distribution of sampling activity in recent decades, over all major groups of biota. A tendency of large numbers of sites to congregate near heavily populated areas is a joint reflection of numbers of available biologists and the tendency to study well-known groups with a relatively local presence.

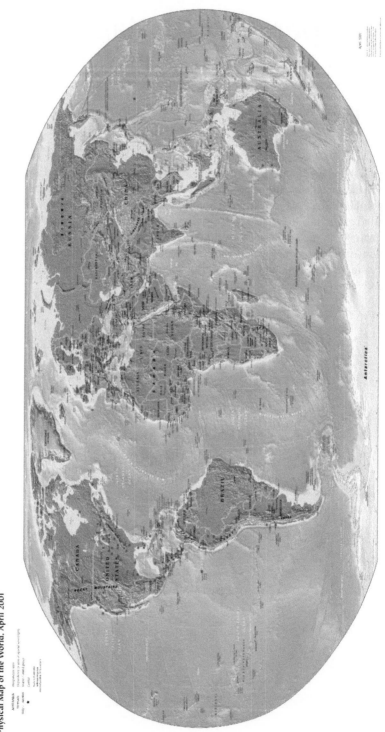

Physical Map of the World, April 2001

Index

A

Abundance
  cyclic changes in abundance, 106–107
  fluctuations in, 103–104
  logarithmic abundance, 48–49
  maximum abundance, 57–58
  randomness in abundance changes, 10
  rank abundance, 26–27, 47–48
Accum program, 84
Accumulation curve, 79
Akaike information criterion (AIC), 11
  maximum likelihood estimator, 11
Arcsine law, 105

B

Berger-Parker index, 38
BestFit program, 69, 162
Biodiversity
  Berger-Parker index, 40
  biodiversity array, 41
  Brillouin index, 38
  broken stick model, 8
  Margelef index, 38
  Menhinick index, 38
  Shannon index, 37
  Simpson index, 37

C

Canonical sequence, 19, 24–26
  derivation of canonical formula, 152–153
Canon utility program, 163
Capacity constant of J distribution, 20–21
Chi-square test, 49–52
  ChiSquare statistical program, 162
  positive test using multisample data, 12–14
  rule-of-five, 162
Chi-square theory, key feature of, 13
  chi-square theorem, 109–113
Computer research tools, 161–163
Counterexamples, 59–61, 65–66
Cyclic changes in abundance, 106–107

D

Density regulated populations, 15

E

Effective randomness, 104
Episodic speciation, 130–132
Epsilon (parameter)
  derivation of, 148–149
Error of misplaced generality, 10
Error terms, behavior of, 73–74
Exact ecology, 57
Exponential distribution and time series, 154–155
Extended sampling, 79–90
  accumulation curves, 79, 81–87
  effect of sampling on parameters, 79–81
Extinction
  extinction rates in the MSL system, 133–134
  extinction and speciation in natural and
    artificial communities, 132–135

F

Field data, compiling and analyzing of, 45–55
Field methods and theory development, guide to,
  140–141
Fisher-Corbet-Williams method (prediction of
  species richness), 61–63
Fossil J-curves, 121–135
  stochastic genera, 129

G

Galton-Watson process (in abundance
  fluctuations), 123
General theory of sampling, 42–44
Goodman statistic, 63–64

H

Hudson's Bay trapping data, 15
Hurlburt's formula, 84–85
Hyperbolic equilibrium theorem, 145
Hyperbolic formula (J distribution), 85–87
  resemblance to power law, 89

J

Jackknife estimator, 64–65
J-curve and J distribution, 1–17; *see also* fossil
  J-curves

J distribution
    density function for, 19–22
    derivation of for stochastic systems, 145–149
    detection of in natural populations, 99–102
    discrete version of, 22–24
    effect of log transformation on the
        J distribution, 28–29
    estimating parameters of, 49–53
    evolutionary origin of, 128–132
    first appearance of, 4–6
    in taxonomic data, 123–128, 171–174
    logistic-J distribution, 1
    mean and variance of the J distribution,
        21–22, 151–152
    presence of in taxonomic data, 123–128
    species and individuals, 22–29
J distribution and its variations, 19–29

K

Kolmogorov-Smirnov (K-S) test, 52–53

L

Linear congruential generator, 32
"Little Ice Age", 132
Logarithmic abundance, 48–49
Logistic-J distribution, 1
Log-series distribution, 155–156
Lotka-Volterra equations, 15, 107

M

Margelef index, 38
Mathematical notes and computer tools, 145–163
Mean and variance of the J distribution, 21–22
Menhinick index, 38
Meta-study
    central methodology, 111–113
    chi-square theorem and test in, 109–113
    converting and compiling scores of, 117–118
    data collection procedure for, 116–117
    illustration of multiple-source histograms,
        applying chi-square theory to, 113–116
    results of, 126–128
    score conversion process in, 114–116
MLS program, 161
Multispecies logistic system (MLS), 93–95,
    161
    generalizations of, 93–95

O

Overlap utility program, 163

P

Perturbation of samples, 111
Pielou transform 1, 42, 141
    application of, 34–40
Populations
    decline in, 75–77
    density regulated populations, 15
    fluctuating populations, 14–17
    predator-prey cycles, 15
Power law, 89, 122
Power series summation of, 150–151
Predator-prey cycles, 15

R

Racetrack analogy, 16–17
Random hierarchy, proof of, 154
Random number generator, 32
Random walks
    on abundance axis, 95
    experiment with, 105
    models of, 104
    population datasets with, 15
    species abundance as, 102
    system behavior resembling, 92
Randomness
    effective randomness, 104
    generating random numbers, 32–33
    pseudorandom generator, 32
    randomness and random numbers, 31–33
Rank abundance, 47–48
    implicit formula for, 26–27
Rarefaction, 79, 81
Research tools, stochastic systems as, 143–144
Richness estimation, 171
    bootstrap method, 65
    estimation methods, 58–66
    Fisher-Corbet-Williams method (prediction
        of species richness), 61–63
    Goodman statistic, 63–64
    inadequacies in current methods, 61–65
    jackknife estimator, 64–65

S

Sample intensity, estimation of, 36–37
SampleSim program, 71
Sampling
    perturbation of samples, 111
    sample simulation algorithm, 41–42
    similarity index of samples, 53–54
    variety of sampling activity, 35–6
Sampling and estimation programs, 162
    general theory of, 42–44

Seasonal probability curve, 107
Speciation
    episodic speciation, 130–132
    speciation in stochastic communities,
        134–135
Species richness, 58–66
Stochastic behavior
    arcsine law, 105
    long runs in, 104–106
    seasonal probability curve, 107
    stochastic abundances, 95–97
    stochastic orbit and its variations, 102–104
    stochastic vibrations, 96
Stochastic communities in nature, 97–107
Stochastic systems
    as research tools, 143–144
    compartmentalized trophic system, 144
    definition of, 93
    research tools, stochastic systems as,
        143–144
    seasonal stochastic system, 144
    weakly stochastic system, 94, 143
    stochastic systems and the stochastic
        community, 91–107
Stochastic species
    communities, 97–99
    null hypothesis, 12–13, 50, 141
Stochastic species hypothesis, J distribution and,
    2–4
Stochastic orbit and its variations, 102–104

T

Taxonomic data, testing for J distribution in,
    171–174
Theory and open problems, summary of, 137–144
Transfer equations, 156–158
Tree of Life Web Project, 125

U

Univoltine distribution, sampling of, 159–161
Utility programs, 162–163
    Canon, 163
    HGen, 163
    Interpol, 163
    KSTest, 162
    Overlap, 163
    SampleSim, 71
    ScanFit, 162
    SolveIt1, 162
    SolveIt2, 163
    StoComm, 161

V

"Veil line" concept, fallacy in, 141

W

Weierstrass Uniform Approximation Theorem,
    43–44